T0338448

OPTICAL SIGNAL PROCESSING

edited
by

M. A. Fiddy
University of Massachusetts, Lowell

M. Nieto-Vesperinas
Instituto de Óptica

A Special Issue of
MULTIDIMENSIONAL SYSTEMS AND
SIGNAL PROCESSING
An International Journal
Vol. 2, No. 4 (1991)

KLUWER ACADEMIC PUBLISHERS
Boston/Dordrecht/London

**THE KLUWER INTERNATIONAL SERIES
IN ENGINEERING AND COMPUTER SCIENCE**

Contents

Special Issue: Otpical Signal Processing
Guest Editors: M. A. Fiddy and M. Nieto-Vesperinas

Distributors for North America:
Kluwer Academic Publishers
101 Philip Drive
Assinippi Park
Norwell, Massachusetts 02061 USA

Distributors for all other countries:
Kluwer Academic Publishers Group
Distribution Centre
Post Office Box 322
3300 AH Dordrecht, THE NETHERLANDS

Library of Congress Cataloging-in-Publication Data

Optical signal processing / by editors, M. A. Fiddy, M. Nieto
    -Vesperinas.
        p.   cm.--(The Kluwer international series in engineering and
    computer science ; SECS 153)
            ISBN 0-7923-9215-9
        1. Optical data processing.   2. Signal processing.   I. Fiddy, M.
    A.   II. Nieto-Vesperinas, M.   III. Series.
    TA1632.0674   1991
        621.36'7--dc20   91-26732
                                                                             CIP

Copyright © 1992 by Kluwer Academic Publishers

All rights reserved. No part of this publication may be reproduced, stored in a retrieval
system or transmitted in any form or by any means, mechanical, photo-copying, recording,
or otherwise, without the prior written permission of the publisher, Kluwer Academic
Publishers, 101 Philip Drive, Assinippi Park, Norwell, Massachusetts 02061.

Printed on acid-free paper.

Printed in the United States of America

Multidimensional Systems and Signal Processing 2, 355 (1991)
© 1991 Kluwer Academic Publishers, Boston. Manufactured in The Netherlands.

Editorial

It gives me great pleasure to credit Dr. M.A. Fiddy and Dr. M. Nieto-Vesperinas for organizing the first Special Issue for this journal. In a relatively short span of time, they were able to procure contributions from researchers active in the area of optics and have them reviewed and processed for the readers to enjoy a very timely issue. It is indeed satisfying to see the links between optics and multidimensional processing underscored in this Special Issue. The exchange of information between scientists of different but related backgrounds, which this will promote, serves very well one of the key objectives of this journal. I am sure that Drs. Fiddy and Nieto-Vesperinas will welcome comments from readers on their worthwhile effort.

N.K. Bose
Editor-in-Chief

Multidimensional Systems and Signal Processing 2, 357-358 (1991)
© 1991 Kluwer Academic Publishers, Boston. Manufactured in The Netherlands.

Introduction

Special issue on Optical Signal Processing

At the suggestion of the Editor-in-Chief, Dr. N.K. Bose, we invited various researchers to contribute to a special issue which focused on the use of optical hardware and methods to solve multidimensional signal processing tasks. This special issue contains papers which span a range of applications of optics to signal processing. Optics here refers to the imposition of information on a two dimensional wavefront and the modulation of this information by the optical processor. The intrinsic parallelism of optics suggests very high computational throughputs provided the information to be processed can be read onto and read from the propagating wavefront sufficiently rapidly. The *first paper*, by Fiddy, entitled "Multidimensional Processing: Nonlinear Optics and Computing," describes some of the background to the use of optics in this context. It is primarily written for the nonspecialist in optical processing and reviews some of the potential advantages and prospects for optical processing and computing. This tutorial paper discusses both the needs of computing and the developing optical hardware and materials required. The *second paper*, by John Caulfield, takes a very fundamental look at the "in principle" advantages one can expect from an optical processing system. It is entitled "Space-time Complexity in Optical Computing" and describes how the spatial and temporal complexity of the computing hardware is related to the complexity of the problem. In particular, the possibility of performing a "fan-in" of data optically leads to computational advantages not realizable by nonoptical means.

One of the earliest demonstrations of optical computing was the use of the simple convex lens to perform a Fourier transformation. Much has been written about Fourier optical correlators and in many ways they represent a mature branch of optical computing. In practice, the robustness of such Fourier optical techniques is still limited because of sensitivity to variations in scale and rotation of the feature to be recognized, with respect to its appearance in an input image. The paper by Mendlovic et al., entitled "Composite Reference Image for Joint Transform Correlator," shows how a complex reference image and a composite Fourier planse filter can improve the selectivity of a real time optical correlator of this kind.

The paper by Pantelic considers a specific operation that is computationally intensive but which can be performed relatively straightforwardly using optical hardware. The paper is on "Optical Computation of Sector and Radon Transforms using a Pinhole Array." An optical system is described which computed the sector transform in a fully parallel fashion. Sector transforms are useful for noise insensitive pattern recognition but are computationally time consuming. A large number of optical computing applications are described in the paper by Kitayama and Ito. This paper on "Optical Signal Processing using Photorefractive Effect," outlines many novel applications such as logic operations, optical storage and neural networks which can be performed optically. Their work focuses on the use of the photorefractive materials which have aroused a great deal of interest in the optical processing

3

community over the last ten years. These materials have a refractive index which changes as a function of the illuminating intensity and thus permits information to be stored, modulated or amplified in near real time. Many novel applications in optical signal processing are proposed and demonstrated.

Finally the paper by Navarro and Tabernero entitled "Gaussian Wavelet Transform: Two Alternative Fast Implementations for Images," describes methods for the efficient encoding of image information in a way that is based on human visual models. Schemes for multi-resolution image coding are described and such complex procedures for the efficient compression of data are good candidates for parallel optical processing, as and when that hardware advances to the required degree of sophistication.

While this special issue in no way encompasses the entire field of optical computing, it does provide an insight into some of the issues of optical processing and presents some of the latest developments in this fast moving field.

M.A. Fiddy and M. Nieto-Vesperinas

Multidimensional Systems and Signal Processing, 2, 359–372 (1991)
© 1991 Kluwer Academic Publishers, Boston. Manufactured in The Netherlands.

Multidimensional Processing: Nonlinear Optics and Computing

M.A. FIDDY

Department of Electrical Engineering, University of Massachusetts at Lowell, Lowell, MA 01854

Received December 30, 1990, Accepted March 20, 1991

Abstract. The purpose of this paper is to overview some of the trends and directions in computing, as performed by optical hardware, resulting from the demands made by multidimensional signal processing. Optical informa- tion processing or optical computing is a vast field and some of the more significant issues are discussed here. We discuss future developments and architectural consequences for such potentially highly parallel and intercon- nected processing systems. Particular emphasis is placed on energy and speed considerations, associated with the use of nonlinear optical materials in optical systems and devices.

1. Introduction

In this paper, we discuss computing and complexity in computing, with optical hardware implementations in mind. The demand for new and continually improved computing capa- bilities increases constantly. Optics is frequently presented as providing a technology for new types of highly parallel and complex computing architectures. Some very general issues concerning complexity and the potential advantages of optical computing are addressed in the accompanying paper by Caulfield. We review the advantages and disadvantages of optical processing and describe the requirements for optical components used in switching and beam modulation. Finally, trends in the development of the subject will be considered, including the increasing attention being paid to new classes of nonlinear optical materials. This brief paper is by no means complete, but reflects a personal view of some of the key issues in the subject. There are many excellent reviews and books written on the subject, just a few of which are cited here, [Horner 1987; Mandel, Smith and Wherrett 1987; Neff 1987; Goutzoulis 1988; Feitelson 1988; Arrathoon 1989; Wherrett and Tooley 1989; Arsenault, Szoplik and Macukow 1989; Reynolds 1989; Caulfield and Gheen 1989; Berra 1989; Optics News 1986; Opt. Eng. 1986; see also special issues in Opt. Eng. 1989; OSA Technical Digest Series 1989; Opt. Eng. 1990; App. Opt. 1990; Int. J. Optoelectronics 1990].

Optical processing research can be broadly divided into two areas. In one area, attention is focused on optical components that improve the speed and performance of existing com- puting hardware. The increasing speed requirement combined with demands for smaller scale integration and higher interconnect complexity makes developments in optical technology seem natural and evolutionary. The activity is stimulated to improve and build on existing digital electronic computing concepts. As a result, one can expect more power- ful and versatile computing resources developed in a way that advances in technology will be transparent to the user. The other area is focused on carrying out those tasks that are

5

peculiarly suited to optics and which electronics does poorly. It is in the domain of both high speed and high parallelism that the latter is focused. Special purpose hardware that falls into this category might perform two-dimensional correlations or convolutions, solve optimization problems perhaps based on learning principles attributed to artificial neural network models, or simple two-dimensional correlation and Fourier-based computations. This second area suggests the idea of wavefront based computing, which may no longer be based on binary computations but include multivalued, fuzzy or analog signal representations; in analysis, continuum modelling has many advantages over discrete modelling. In each area, and perhaps spanning both, practical necessities lead to a philosophy of combining the best features of optical and electronic hardware through the development of so-called hybrid systems.

2. Computing and Work

If one asks the questions, what is computing, it could be summarized as the transfer and manipulation of information, e.g., of strings of bits of information having some logical meaning. Each component or basic element of a computing system usually has one of two states and computation arises from a sequence of state changes. It is a process that requires some energy to be expended. This can be illustrated through the example of the original mechanical computer of Babbage, developed in the 1840s. One could ask the question whether or not this computer could run backwards. Clearly it could not as there must be friction present in order for it to operate reliably. As a result, it's operation requires energy and will generate heat. Modern computers also generate heat, somewhere of the order of 10^8 kT per logic operation, such as an AND or an ADD on single bytes. One would like to operate at lower energies and also at higher speeds; operating on smaller physical scales can assist with these goals, but there will always exist fundamental physical limitations on the minimum dimensions and energies required. Much work has been done to identify exactly where this energy for computation is needed. For example, in a binary processor, binary digits can be materialized as magnetic domains; there is no heat dissipation if the domain is left alone and energy is required only if the state is to be changed.

We can speculate what the lowest energy might be to represent a binary state. A particle spin-state, for example, might suffice and this could perhaps be altered without dissipation of energy. If one considers an electron in a potential well, its probability of escape is $\approx$ $\exp(-\Delta E/kT)$ where $\Delta E \approx kT = 4 \times 10^{-21} J$. This represents a possible limitation on the minimum energy required in computing where bound or unbound electrons distinguish one *state* from another. However, according to Heisenberg's uncertainty principle, if $\Delta E \Delta t \leq h/2\pi$ then quantum mechanical tunnelling may take place, leading to a loss of information. Based on thermal considerations, it is generally accepted that to erase a bit of information requires dissipation of energy $kT\ln2$, (Landauer's principle). One can compare the energy requirements of different devices used in computing. Today's electronic devices require $\approx 10^6$ kT to switch states, and integrated circuits require $\approx 10^{10}$ kT which is of the order of 10^{11} operations/sec/W, if we define an operation as a bit change. A typical mainframe such as a VAX requires 3 $kW \approx 10^{24}$ kT/sec or 350 instructions/sec/W, where an instruction is defined as a single accept and execute step; this translates to $\approx 10^{-4}$ J/bit.

The situation in optics is quite different. Let us assume we require ≥ 10 photons to distinguish an on-state from an off-state. This corresponds to $\approx 10^{-17}J$ and if assume that we need ≥ 100 photons detected for 10 bits, this is $\approx 10^{-16}J$. If one imagines 1000 by 1000 elements measuring at 10 bit accuracy, then this would require $10^{-10}J$ of energy on this basis. One can now compare this requirement with the digital electronic energy requirements of $\approx 10^{6}kt/bit$ for a device to switch or a total of $10^{-8}J$. Optics thus has an energy advantage, in principle. Moreover, one can project that analog optical processors offer about 100:1 decrease in processor volume along with at least a 10:1 decrease in power requirements over their digital electronic counterparts.

Speed considerations are also important. A transistor may switch at ≈ 5 ps, a logic gate at ≈ 120 ps, a chip may have a clock cycle of ≈ 1 ns and a system a clock cycle of ≈ 5 ns. The difference between the speed capability of a component and the system is a factor of 1000. The reason for this is electromagnetic interference, connection complexity and impedance matching effects. In optics, these delaying factors can be reduced since one can accomplish a high degree of interconnectivity by refractive or diffractive elements. For example, one can regard a simple convex lens, focusing a plane wave to a spot, as similar to a high density fan-in element taking light from an extended array of locations to an isolated site, via free space. This potentially high degree of interconnectivity can be achieved with no interference or cross-talk between connections, provided the propagation medium is linear. If one wanted signals to mix or interfere, they could be propagated through a nonlinear optical medium.

The information carrying capacity of optical fields is usually expressed in terms of numbers of channels passing through a given area. It has been shown [Ozaktas and Goodman 1990] that optical channels can be regarded as solid wires with a minimum cross section of $\lambda^2/2\tau$, where λ is the optical wavelength. A consequence of this analysis is that any number of independent wavefields are permitted to overlap in real (image) space or in Fourier space, but not in both. Also, one can deduce from a degrees of freedom argument, that the size or volume of an optical processor is linear in the total communication length. This follows from the fact that a diffraction limited spot does not increase in size on cascading several systems together; i.e., the effective cross-section of each optical channel is independent of length. Hence, with increasing system size, the communication volume required for establishing optical interconnections will grow more slowly than that required for establishing conductor-guided interconnections, [Ozaktas and Goodman 1990].

3. Computing Architectures

Conventional digital electronic processing relies on a serial processing architecture which leads to an input-output bottle neck. This bottle-neck manifests itself in fundamental limitations on computing speed since all processing has to be performed by passing information through a single cpu. Obviously present day systems exploit parallel or multiprocessor architectures, [Hwang 1987], and these are referred to as fine-grained or coarse-grained processors, operating either in a single instruction multiple tasks or multi-instruction multiple-task modes, which alleviates the bottleneck to some extent.

Recently there has been interest in so called artificial neural processing architectures, [Khanna 1990]. These architectures are very loosely based on a simple model for the brain and assume a high degree of interconnectivity between sets of simple processing elements such as two-state switches. The power of this kind of computing resides in the connection strengths between processing elements. These connection weights can be modified and so the performance of the network can be influenced by a learning rather than algorithmic rule. The expectation is that such a network will be able to more easily solve recognition and optimization problems of the kind that are currently difficult to solve by conventional processing means, [Soffer, et al. 1986; Caulfield, Kinser and Rogers 1989; Owechko 1989; Psaltis 1990].

A system consisting of many processing elements, each one of which makes a weighted connected to all others, is referred to as a fully connected architecture. If a fully connected parallel processing hardware were available, there are several problems that could be programmed to be solved on it, [e.g., Abbiss, Brames, Byrne and Fiddy 1990; Steriti, Coleman and Fiddy 1990]. Since the key requirement is a highly interconnected set of simple processors, this kind of architecture has been of considerable interest to the optics community, since it looks likely that optical hardware could more easily provide the large parallel systems necessary. Optically, a *processing element* may take the form of a bistable optical switch which either transmits or does not transmit light, according to the light level incident on it. It is still early days in hardware development of neural computers, but there have been many proposed optical architectures which look promising, once the necessary optical components become available, [Caulfield, Kinser and Rogers 1989].

If one accepts that there could be advantages in processing information with photons rather than electrons, one has to identify classes of materials with the performance required for this technology to be competitive. Higher speeds, higher interconnectivities and lower overall energy requirements are motivating conditions for this approach, [Feldman, et al. 1988]. However, electronics will continue to miniaturize further, with associated reductions in processing speeds and energy requirements. The fundamental limitations associated with free space optical interconnections, resulting from the constant radiance theorem [Goodman 1985] and recently discussed in the context of tubes of information by Ozaktas and Goodman [1990], make it unlikely that the physical dimensions of optical components will be much different to those used in digital electronic devices. Diffraction limitations will ultimately restrict the density of cross-talk-free interconnects that can be experimentally realized. The advantages optics can offer will have to be found in the ease of generating high density but nonwired interconnects and the prospects for low energy operation.

4. Spatial Light Modulators

A major bottleneck in optical processing has been that of imparting the required information onto the optical beams. Sources and light modulators have their own intrinsic limitations in terms of speeds, resolutions and efficiencies, and at present these spatial light modulators constitute a weak link in any optical computing architecture. There are many mechanisms that can be exploited for this purpose, such as electro-optic, magneto-optic, acousto-optic, photorefractive, mechanical (e.g., deformable mirrors) etc., [special issue of App. Opt. 1989; Neff, Athale and Lee 1990].

Perhaps the two most commonly available SLMs are liquid crystal based or magneto-optic. Commercially available SLMs such as liquid crystal based *light valves*, tend to have 2 to 100 ms response times, 20 to 100:1 contrast ratios and 16 to 30 line pairs/mm resolution. Both optical and electronic addressing is possible. Magneto-optic devices tend to have response times in the tens of nanosecond range and frame rates ranging from 350 to 1000 Hz, with high contrast ratios (10,000:1) but low resolutions of the order of 10 line pairs/mm. The magneto-optic pixelated devices are binary in character and nonvolatile, with up to 256 by 256 pixels available, [special issue of App. Opt. 1989].

Newer mechanisms for SLMs suitable for digital optical computing are being developed. For example, in so-called SEED (self-electro-optic-effect) devices, excitonic absorption in a layer of quantum wells can be controlled by an applied voltage. These offer 2 ns switching speeds and 5:1 contrast ratios, and arrays of 32×64 elements or 50 line pairs/mm resolutions, [Miller et al. 1985]. The symmetric or S-SEED switches state if about 1 pJ of light is incident on it; they cost $\approx$ \$14k per array. Such multiple quantum well materials also provide potentially low threshold lasers and hence modulating sources. Their low threshold of operation arises because as the volume of the gain medium is reduced the material losses are reduced and as the reflectivity of cavity mirrors is increased, the cavity loss reduces further.

New materials are appearing routinely, which may make significant improvements to the SLM situation. We cite, for example, erasable dye polymers and polymers which can be addressed in the infrared and undergo phase changes [e.g., Roland 1990]. New components such as arrays of surface-emitting microlasers can be used for computational purposes, either for logical functions or interconnections.

5. Computing Needs

There are some areas of computing into which optics is already providing improvements. We cite for example communications, e.g., via fiber optics. The research bit rate limit today is around 350 Gbits/sec with a colliding pulse mode-locked laser and predictions of 1 Tbit/sec by the year 2000. Currently, up to 1 Gbit/sec communications between chips has been reported (MIT Lincoln Laboratories) and guided wave or free space interconnects promise further improvements; see for example the use of overhead holograms relaying clock pulses across VLSI chips, [Goodman et al. 1984; Kostuk, Goodman and Hesselink 1985]. Problems with wire interconnects arise when their length is limited to approximately that of the wavelength of the signal and at higher frequencies, shorter interconnect lengths are required. Currently, computer speeds are limited by interconnect times (100 ps) rather than switching times of devices (10 ps for GaAs devices). Optics should help overcome clock skew problems, ground loop isolation, cross-talk (which increases at higher frequencies), losses and impedance matching problems. Also, 3-D transmission is possible with optics between arbitrary locations on boards.

Another computing need is data storage. Magnetic media are capable of 0.15 Mbits/sq.mm or 100 MBbits/sq.in. with IBM having announced 1 Gbit/sq.in.; the storage capacity for magnetic media appears to double every 2.5 years. For optical storage, such as magneto-optic or phase change storage, 645 Mbit/5.25 in. disk is virtually the entry level storage density, with further increases in density expected. An advantage of optical disk storage is the fact that the recording head need not be too close to the storage medium, minimizing problems of head crashes, etc.

A number of materials have been proposed for optical storage and/or reconfigurable interconnects and the more promising ones are shown in Table 1.

High capacity storage is possible in a volume rather than disk media. Also, volume or Bragg diffraction results in high efficiency optical elements. Most volume storage concepts are based on the idea of holographic storage, whereby readout is by addressing the medium with a specific *reconstruction* wave, or read-out is by association. The latter is frequently equated to a neural network content addressable or associative memory, in which partial information acts as the key to read-out the entire data set.

The storage capacity theoretically possible in a volume medium is given by $N = V/\lambda^3$ bits, which results in $N = 10^{12}/cm^3$ assuming $\lambda = 1\ \mu$ m. In practice, the number of stored bits is determined by such considerations as desired diffraction efficiency, crosstalk, etc. For example, to obtain equal diffraction efficiency for angularly multiplexed (i.e., reference waves with different incident angles) holograms, an elaborate exposure scheduling technique has to be devised in which successively written holograms take into account the previously written holograms. Calculations suggest that about 100 holograms with 1% diffraction efficiency can be multiplexed in a volume medium. However, increasing the number of multiplexed holograms greatly increases the precision with which the exposure time and laser power must be controlled; typically for 100 holograms the recording energy must be maintained to better than 0.1%.

Given a storage capability in excess of 1000 lines/mm, one can expect to write approximately 0.25 Mbits/sq.mm. (160 MBits/sq.in.) in principle. Thus, a target storage capacity of 10^{12} bits stored would require a disk area of 4×10^6 mm^2, which is a total of 4 square meters. However, if volume storage were employed, and an effective resolution capability of 1.25×10^8 bits/mm^3 is possible, which it is in principle, this would reduce the required storage volume to 8×10^4 mm^3, which is a volume of just 80 cubic centimeters or approximately 4.5 cm-cube. With smaller wavelengths, improved coding and track squeezing, 8 G bits/sq. in. is projected.

Table 1. Materials for storage and interconnects.

Physical effect	Typical materials	Resolution lines/mm	Response time/sec	Sensitivity mJ/cm^2	Diffraction efficiency
Photorefractive	BSO, BGO SBN	≈ 2000	10^{-3} to 1	100	10%
Thermoplastic	Stabelite ester + PVK	1000	0.1 to 1	0.1	20%
Magneto-optic	GaTbFe	1000	10^{-7}	100	$< 10^{-3}\%$
Photo-polymers	Dyes in polymers	>1000	10^{-6}	50	20%
Deformable polymers	Elastomers	1000	10^{-8}	30	7 to 10%
Phase change	InSbTe	1000	10^{-7}	100	—

6. Optical Computing

The subject of optical computing *per se* has been studied since the early 1960s, [Goodman 1989]. Initially processing was carried out through the use of convex lenses which can be shown to perform a Fourier transformation on a wavefront, [Reynolds et al. 1989]. Since the Fourier transform plays such a key role in many signal and image processing applications, the idea of performing this operation in real time directly on an image-bearing wavefront was clearly appealing. Many designs for optical processors, i.e., analog optical computers, based on this simple principle have been considered over the years, but few hardware implementations have been incorporated into systems. Exceptions are optical processors for the processing of synthetic aperture radar images, [Cutrona et al. 1966], and compact optical correlators designed primarily for military use, [Flannery and Horner 1989; Caulfield et al. 1987; Gregory and Kirsch 1988]. One of the factors reducing the impact of optical processors of this type has been the need for a possibly expensive SLM for input of information. Another reason has been the belief that accuracy would invariably be limited to about 4 bits which was not considered satisfactory for many applications, especially if further digital image processing was required. Also, of course, the reducing costs of digital electronic hardware to perform similar functions, did little to stimulate optical solutions.

It is still essentially true today that analog processing, while recognized as being a powerful and an elegant solution to many multidimensional processing problems, is not widely pursued. However, further study of these better established analog optical technologies is finding some encouragement these days, for example through the DARPA TOPS (Transition of Optical Processors to Systems) program. Optical procedures will be developed in this program for channelizers, pulse compressors, real time synthetic aperture radars, null steering, pattern recognition, optical control of phased arrays, precision direction finding and data base management.

The question remains unanswered as to whether digital optical computers can perform better than their digital electronic counterparts. Much has been written specifically on digital optical computing, [Jenkins et al. 1984; Jenkins 1984; Prise, Striebl and Downs 1988; Striebl et al. 1989; Cathey, Wagner and Miceli 1989]. AT&T has announced a functioning prototype digital optical computer, [Huang 1990]. It employs four arrays of S-SEEDS, using only 32 of the 2048 S-SEEDS on each chip; each S-SEED acts as a NOR logic gate. The prototype systems operates at one million cycles per second and has a potential speed of one billion switching operations per second. Hughes Research Labs have developed an acousto-optic [Lee and Vanderlugt 1989] matrix processor referred to as PRIMO (Programmable Real-time Incoherent Matrix Optical Processor), [Owechko and Soffer 1989]. It can perform 10 billion multiplications and additions/s and occupies a small volume. It is currently based on a 256 by 256 SLM and can be used as a Fourier processor and correlator. OptiComp's DOC2 (digital optical computer 2) comprises 64 lasers and exploits acousto-optic devices to impart data onto a beam. This system, in principle, has the potential to operate at one trillion binary operations per second, but is currently limited in performance by the electronic computer which provides instructions to the optical processor. Guilfoyle, founder of OptiComp, predicts the capability of searching 10,000 pages of text a second and running the system on only 200W of power. Our own group has worked in close collaboration with Semetex Corporation on the construction and evaluation of a digital optical processor based on the Semetex magneto-optic Sight-Mod. Using cascaded Sight-Mods, with binary data sets input to each, one can interpret the multivalued output levels directly in terms of Boolean operations between these data sets. Performance is currently limited by the fact that the processor is driven by a serial computer, but the optical branch, with just two cascaded 256 by 256 SLMs, is capable of 12.2 million operations per second.

This general question about whether optical processors can out perform electronic processors is the focus of the accompanying paper by Caulfield [1991]. In that paper he argues that if one breaks the solution of a problem down into its time and space complexity, i.e., the number of clock cycles and the number of cpus necesary to compute a solution, then an optical processor has an intrinsic advantage. This advantage lies in the fact that the processor implicitly performs a fan-in or a fan-out by optical means, a process which normally would expend computational time or effort, but which happens in real time, through the use of bulk refractive optics, in an optical processor.

7. Nonlinear Optical Materials

One of the factors that rejuvenated work in optical processing in the late 1970s was the discovery of materials with large third order susceptibilities. These materials provided a mechanism for faster and lower power switching of light by light, in parallel configurations. There are certain fundamental bounds on the peformance of any such bistable switching element, based on a Fabry Perot etalon structure containing a nonlinear medium between its reflecting surfaces. These limits were first suggested by Smith [1982]. Constraints on performance involve limitations in heat dissipation that might be anticipated (e.g., 100 W/cm^2, the thermal transfer limit), a quantum (tunneling) limit and a thermal limit (electron confined in potential box). Theses various considerations lead to an expected optimal performance around 0.1 μW/bit being possible, in principle, for switching at 0.1 ps. This specification exceeds that of all current optical devices, (see also [Smith 1982; Keyes 1985]). It is instructive to note that the human neuron performs with switching times of tens of ms and power requirements of nW per bit.

All optical materials exhibit some degree of nonlinearity in the sense that their refractive index is a function of an externally applied voltage or is a function if the incident light intensity. Materials with a refractive index dependent on the light intensity are known as Kerr media or χ^3 media and all materials fall into this category. Those whose index is a function of the light amplitude or an externally applied voltage are known as Pockel's media, (e.g., [Hopf and Stegeman 1986]). We consider here only the role of Kerr-type media for optical switching, but do so for illustrative purposes and in no way mean to dismiss the importance of χ^2 media. Table 2 illustrates the range of values possible for these nonlinear media.

These materials could be used in devices like the nonlinear Fabry Perot etalon or a Fredkin gate, [Shamir et al. 1986]. The mechanisms responsible for the nonlinearity depend upon the size and material properties of the medium. Mechanisms include thermal effects, which might be low, electrostrictive forces and local field effects, real transitions, i.e., absorption and resonance phenomena and virtual transitions (no losses) such as electronic nonlinearities, which can be very fast. Nonlinear polarizability effects could be as fast as a femtosecond, while the large χ^3 associated with fluorescein doped Boric acid glass may take 10 seconds.

Table 2. Examples of values of χ^3.

$$D = \epsilon_0 E + P = \epsilon_0 (1 + \chi) E = \epsilon_0 n^2 E$$

$$P = \epsilon_0 \{\chi^1 E + \chi^2 E^2 + \chi^3 E^3 \ldots\}$$

We consider only third order nonlinearities here, hence

$$n = n_0 + n_2 |E|^2 \text{ and } n_2 \text{ in m}^2/\text{W} \approx (4\pi)^2 \, 10^{-7} \, \chi^3/3 \text{ in esu, i.e., 1 esu} \approx 1 \text{ cm}^2/\text{kW}$$

χ^3 esu	
10^1	fluorescein doped Boric acid glass [Tompkin, Malcuit and Boyd 1990]
10^0	fluorescein absorbed to gold spheres InSb at 77 °K
10^{-1}	$Hg_{1-x} CD_x Te$ (at resonance)
10^{-2}	microparticles and quantum confinement: close to resonance [Hache, Ricard and Flytzanis 1986]
10^{-3}	thermal results with colloidal gold [Lai, Leon, Lin and Fiddy submitted] and electrostriction theory for coated particles [Neeves and Birnboim 1989]
10^{-4}	various thermal effects
10^{-5}	chinese tea [Zhang et al. 1989]
10^{-6}	vanadium pentoxide rods
10^{-7}	gold microparticles in glass; Fermi smearing theory [Bloemer, Haus and Ashley 1990] 0.5 μm glass particles in water [Smith et al. 1981]
10^{-8}	colloidal gold and nanocrystals in glass
10^{-9}	0.2 μm latex spheres; quantum confinement: theory
10^{-10}	Ge, Si, GaAs induced polarization
10^{-11}	
10^{-12}	CS_2 molecular reorientation
10^{-13}	most liquids and glasses

In an etalon, as the incident optical field increases in intensity, the refractive index of the medium between the mirrors increases, shifting the transmission peaks of the etalon to other wavelengths. The speed of the device is determined by the build up time of the resonator; thus one can reduce the response time by reducing the length of the cavity, but then this, of course, requires that more power be used to induce the same refractive index change. Such a nonlinear etalon can function as a simple Boolean logic gate and, because of local field effects leading to bistable behavior, can function as a latching device also.

The Fredkin gate was originally described by Bennett [1973], who argued that dissipation in such a system could be arbitrarily low if computations were carried out in a thermodynamically reversible fashion. The basic Fredkin gate consists of a simple Mach-Zehnder interferometer, in which two incident information bearing waves are split to each pass both ways around the interferometer. In one arm of the interferometer, there is a nonlinear optical medium whose properties can be controlled by a third beam, not passing around the interferometer path. When the third beam is on, the two input beams are reversed at the output. Milburn [1989] argued that this architecture might allow switching in a reversible and error free way, provided a lossless nonlinearity was used to produce intensity dependent phase shifts in one arm of the interferometer, and provided only one photon was used at a time. This latter requirement clearly makes the device impractical since an extremely large χ^3 would be required. However, with realistic χ^3, and larger photon numbers, classical field fluctuations lead to phase fluctuations and some degree of error, but maybe an acceptable performance; see also R.W. Keyes [1989].

Optical systems incorporating nonlinear media and a degree of feedback, are capable of exhibiting competitive and cooperative dynamics. Rather than design an optical system to emulate parallel digital hardware one could exploit the unique characteristics of an optical system to perform a wholly new kind of computing. The only requirement for the latter, i.e., cooperative dynamics, is some nonlinear component in the processor and a feedback mechanism. The high degree of interconnectivity possible by wavefront processing and the similarity to neural network processing has not gone unnoticed. The development of optical processing systems is now progressing hand in hand with research into the most useful aspects of artificial neural nets, (see e.g., [Anderson 1990]).

8. Photorefractive Materials

The discussion of optical processing would not be complete without making reference to photorefractive media. Photorefractives can be used to record holograms in real time and this can be exploited in a number of different ways, see for example the accompanying paper by Kitayama and Ito [1991]. They can provide temporary storage media for information recall or for information interconnections. The hologram, as mentioned previously, is a particularly powerful processing concept in computing, since it permits auto- and hetero-association in recall. This kind of recall thus constitutes a data driven or data-base driven processor.

In a photorefractive, two beams interact within the volume of the medium and produce a refractive index distribution that mimics the optical interference pattern, [Gunter and Huignard 1988; Yeh et al. 1989]. A simplified picture of the mechanisms operating in photorefractives is as follows. Trapped charges become excited and move into the conduction band. They then move into relatively dark regions where they become trapped, leading to a variation in the local space charge field. This in turn modifies the refractive index of the material as a result of the large linear electro-optic coefficient of the photorefractive, i.e, a χ^2 process.

Photorefractives have been used for a variety of applications such as storage, linear association and amplification of information. Unfortunately, as the hologram is interrogated, it will be partially erased, unless one can choose a medium which is very slow to respond. A particularly interesting feature of photorefractives is their ability to holographically store information without the explicit need to introduce a reference wave. This phenomenon, known as self-pumped phase conjugation, makes use of a fanning effect that can occur within the photorefractive and which provides a self-referencing set of interfering waves, [Feinberg 1982]. This self-pumped wave has been used in associative memory applications [Owechko 1989], and its stability for storage and efficient recall of information has been widely studied, (e.g., [Lin and Fiddy 1991]).

9. Reversibility and Wave Particle Duality Processors

In recent papers an interesting new concept has been proposed by Caulfield et al. with regard to the computing capability of optical processors, [Caulfield and Shamir 1990;

14

Caulfield, Shamir, Ludman and Greguss 1990]. The kT barrier mentioned earlier is claimed not to be a real barrier, but can be much smaller than this, for the following reasons. One can consider a wavefunction associated with each photon, and one can regard each photon as passing through all possible paths in an optical system. For example, in the classic Young's slit experiment, one can regard each photon as actually passing through both slits, in order to explain the fringe pattern obtained, even when only one photon at a time is present in the system. In this way, on passing through an optical computing architecture, one can regard each photon as performing all the possible interconnections and computations that exist. Until a photon is detected, information is conserved and dissipative measurements are avoided. This interpretation of a photon as an all pervasive wave requires only that the degree of coherence of the radiation is sufficiently high for this assumption to hold throughout the optical system. This is the case with the coherent optical Fourier processor mentioned above.

10. Conclusion

Optics can offer fast switching at low powers and a high degree of parallelism, as well as providing high degrees of fan-in and fan-out. This is the case in principle, but devices that truly exploit these possibilities and can be incorporated into computing systems are still somewhat limited.

If information is to be manipulated many times over, then it must be standardized somehow and continuously amplified in order to be properly preserved, [Keyes 1985]. Unfortunately, optical power drops as $1/N$ and electrical power as $1/N^2$ for 1 to N fanning, [Goodman 1985] so signals may have to be restored by amplification. Once any tranduction of photons back to electrons is performed, speeds and energy requirements degrade, sometimes to the point that the advantages of optical processing are significantly reduced. It is important therefore, for a successful optical processing application, that there is as little interchange of information between electrons and photons and back to electrons, as possible. It was pointed out by Miller, [1989], that interconnecting two high impedance logic devices separated by a very low impedance medium such as space, will require a high impedance transmission line that will inevitably incur high losses. However, optics avoids this problem because optical devices are quantum devices or ideal impedance transformers; a photodiode, for example, matches a low voltage in a low impedance medium to a higher voltage in a higher impedance device. There is therefore a fundamental distinction to be made between trying to measure a voltage and quantum detection, in which one is counting photons.

While there are few proposed optical devices that could be compared directly with something like a transistor, optical technology offers advantages in the areas of interconnects and storage. Storage has already benefitted from progress in optics research and we can expect mass optical storage to play an increasingly important role in computing. As materials research progresses, erasable volume holographic memories should become more feasible, and these will offer much greater data rates and different kinds of processing, such as associative or data-base processing. The high degree of interconnectivity that can, in principle, be realized can be exploited if one accepts high densities of data flow rather than high speeds, necessarily. Future developments in hybrid systems seem inevitable, such as smart pixels, with some electronic processing being performed at each pixel in an array.

Of particular importance to the future of optical processing is the development of improved materials and, in turn, improved SLMs; improvements are necessary in speed, energy requirements and resolution, as well as cost. If one could speculate on an ideal SLM, one might request that it have some 10^6 elements with a frame speed of a few μs and very low switching energy, e.g., a few nJ. It is also desirable that the SLM be stable over long periods of time and provide low attenuation and not be too wavelength dependent. The SLM is the key device for input of information and control of the interconnect capability of optics.

While there is a very large investment in digital electronic hardware development, there is a slow but steady infiltration of optical circuitry and componentry into these electronic systems, [Sluss et al. 1987]. It is thus likely that an all optical processor, while some time away because of materials limitations, will evolve step by step over the next couple of decades. It seems more likely that optical processing will be device rather than system driven in the immediate future. This probably reflects the fact that there is no major commercial need requiring a large processing capability of the kind optics could offer at the present time.

References

Abbiss, J.B., Brames, B.J., Byrne, C.L., and Fiddy, M.A., "Image restoration for a fully connected architecture," *Opt. Lett.*, vol. 15, p. 688, 1990.

App. Opt., 28, issue 22 on Spatial Light Modulators, 1989.

Anderson, D.Z., "Competitive and cooperative dynamics in nonlinear optical circuits," in *An Introduction to Neural and Electronic Networks*, (S.F. Zornetzer, et al. eds.) Academic Press: Reading, MA, pp. 349–362, 1990.

Arrathoon R., *Optical Computing*. Marcel Dekker: New York, 1989.

Arsenault, H., Szoplik, T., and Macukow, B. *Optical Processing and Computing*. Academic Press, San Diego, 1989.

Bennett, *IBM J. Res. Dev.*, vol. 17, p. 525, 1973.

Berra, P.B. et al. "Optics and supercomputing," *Proc. IEEE 77*, p. 1797, 1989.

Bloemer, M.J., Haus, J.W., and Ashley, P.R., "Degenerate four wave mixing in colloidal gold as a function of particle size," *JOSA* B7, p. 790, 1990.

Cathey, W.T., Wagner, K., and Miceli, W.J. "Digital computing with optics," *Proc. IEEE*, 77, 1558, 1989.

Caulfield, H.J. et al. "Optical Correlators," *Photonics Spectra*, pp. 117–121, Dec. 1987.

Caulfield, H.J., Kinser, J., and Rogers, S.K. "Optical neural networks," *Proc. IEEE 77*, p. 1573, 1989.

Caulfield, H.J. and Gheen, G. "Selected papers on Optical Computing," *SPIE Milestone Series*, vol. 1142, 1989.

Caulfield, H.J., and Shamir, J., "Wave-particle-duality processors: characteristics, requirements and applications," *JOSA* A7, p. 1314, 1990.

Caulfield, H.J., Shamir, J., Ludman, J.E. and Greguss, P., "Reversibility and energetics in optical computing," *Opt. Lett.*, vol. 15, p. 912, 1990.

Caulfield, H.J., "Space-time complexity in optical computing," *Multidimensional Systems and Signal Processing*, vol. 2, pp. 373–378, 1991.

Cutrona, L.J., et al. "On the application of coherent processing techniques to synthetic aperture radar," *Proc. IEEE*, 54, p. 1026, 1966.

Feinberg, J., "Self-pumped, continuous-wave phase conjugator using internal reflection," *Opt. Lett.*, vol. 7, p. 486, 1982.

Feitelson, D.G., *Optical Computing*. MIT Press: Cambridge, MA, 1988.

Feldman, M.R. et al. "Comparison between optical and electrical interconnects based on power and speed considerations," *App. Opt.*, vol. 27, p. 1742, 1988.

Flannery, D.L. and J.L. Horner, "Fourier optical signal processors," *Proc. IEEE 77*, p. 1511, 1989.

Goodman, J.W., et al. "Optical interconnects for VLSI systems," *Proc. IEEE 72*, p. 850, 1984.

Goodman, J.W., "Fan-in and fan-out with optical interconnections," *Opt. Acta.*, vol. 32, p. 1489, 1985.

Goodman, J.W., "A short history of the field of optical computing," in (Wherrett, B.S. and F.A.P. Tooley, ed.) *Optical Computing*, Proc. 34th Scottish Universities Summer School in Physics, Edinburgh University Press, pp. 7–21, 1989.

Goutzoulis, A.P., "Digital electronics meets its match," *IEEE Spectrum*, p. 21, August 1988.

Gregory, D.A. and Kirsch, J.C. "Compact optical correlators," *SPIE* 960, p. 66, 1988.

Gunter, P., and Huignard, J.-P., "Photorefractive Materials and Their Applications I & II," Springer-Verlag, vols. 61 and 62, (1988); see also *IEEE* QE-25, issue 3, pp. 312–647, 1989, and also, *JOSA* B7, issue 12, 1990.

Hache, F., Ricard, D., and Flytzanis, C. "Optical nonlinearities of small metal particles: surface mediated resonance and quantum size effects," *JOSA B3*, p. 1647, 1986.

Hopf, F.A. and Stegeman, G.I., "Applied Classical Electrodynamics, Vol. 2: Nonlinear Optics," Wiley, New York, 1986.

Horner, J.L., *Optical Signal Processing*. Academic Press: San Diego, 1987.

Huang, A., "Optical Computer: is concept becoming reality," *SPIE OE Reports*, 1, p. 75, 1990.

Hwang, H., "Parallel processing with supercomputers," *Proc. IEEE* OC-75, p. 1348, 1987.

Int. J. of Optoelectronics, 5, issue 2 on Optical Materials, Devices and Systems for Computing, 1990.

Jenkins, B.K. et al. "Architectural implications of a digital optical processor," *App. Opt.*, 23, p. 3465, 1984.

Jenkins, B.K., "Sequential optical logic implementation," *App. Opt.*, p. 3455, 1984.

Keyes, R.W., "What makes a good computer device?," *Science*, 230, p. 138, 1985.

Keyes, R.W., "Making light work of logic," *Nature*, vol. 340, p. 19, 1989.

Khanna, T., "Foundations of neural networks," Addison-Wesley, Reading, MA, 1990.

Kitayama, K. and Ito, F., "Optical signal processing using photorefractive crystals," *Multidimensional Systems and Signal Processing*, vol. 2, pp. 401–419, 1991.

Kostuk, R.K., Goodman, J.W., and Hesselink, L., "Optical imaging applied to microelectronic chip-to-chip interconnection," *App. Opt.*, 24, p. 2851, 1985.

Lai, H.-S., Leon, R., Lin, F.C. and Fiddy, M.A. "Observations of χ^3 in aqueous suspensions of colloidal gold," submitted to *Opt. Lett.*

Lee, J.N. and VanderLugt, A. "Acoustooptic signal processing and computing," *Proc. IEEE* 77, p. 1528, 1989.

Lin, F.C. and Fiddy, M.A., "Optimization of the self-pumped phase conjugation in $BaTiO_3$ for optical image storage and readout," submitted to *JOSA-B*, December, 1990.

Mandel, P., Smith, S.D., and Wherrett, B., "From optical bistability towards optical computing," North Holland, New York, 1987.

Milburn, *Phys. Rev. Lett.*, vol. 62, p. 2124, 1989.

Miller, D.A.B., "Optics for low-energy communication inside digital processors: quantum detectors, sources, and modulators as efficient impedance converters," *Opt. Lett.*, vol. 14, p. 146, 1989.

Miller, D.A.B., et al. "The quantum well self-electrooptic effect device: optoelectronic bistability and oscillation and self-linearized modulation," *IEEE QE-21*, p. 1462, 1985.

Neff, J.A., "Major initiatives for optical computing," *Opt. Eng.*, vol. 26, p. 2, 1987.

Neff, J.A., Athale, R.A., and Lee, S.H. "Two-dimensional spatial light modulators: a tutorial," *Proc. IEEE* 78, p. 826, 1990.

Neeves, A.E. and Birnboim, M.H., "Composite structures for the enhancement of nonlinear optical susceptibility," *JOSA* B6, p. 787, 1989.

Opt. Eng., vol. 25, issue 1 on Digital Optical Computing, 1986.

"Optical Computing," OSA Technical Digest Series, 9, Salt Lake City, 1989.

Opt. Eng., 28, issue 4 on Optical Computing, 1989.

Optics News, vol. 12, issue 4 on Optical Computing, 1986.

Osaktas, H.M. and Goodman, J.W. "Lower bound for the communication volume required for an optically interconnected array of points," *JOSA A7*, p.2100, 1990.

Owechko, Y., "Self-pumped optical neural networks," *OSA Optical Computing Digest*, Salt Lake City, p. MD4-1, 1989.

Owechko, Y. and Soffer, B.H. "PRIMO: a programmable electrooptic processor," *Proc. ASILOMAR Conf. on Signals Systems and Computers*, p. 297, 1989.

Prise, M.E., Striebl, N., and Downs, M.M. "Optical considerations in the design of digital optical computers," *Opt. and Quant. Electronics*, 20, p, 49, 1988.

Psaltis D., et al. "Holography in artificial neural networks," *Nature*, 343, p. 325, 1990.

Reynolds, G.O., et al. *The Physical Optics Notebook: Tutorials in Fourier Optics.* SPIE Press, Bellingham, 1989.

Roland, C., see "Navy is exploring erasable optical media," *Photonics Spectra,* p. 70, Dec. 1990.

Shamir, J. et al. "Optical computing and the Fredkin gate," *App. Opt.,* 25, p. 1604, 1986.

Sluss, J.J. et al. "An introduction to integrated optics for computing," *IEEE Computer,* p. 9, Dec. 1987.

Smith, P.W. et al. "Four wave mixing in an artificial Kerr medium," *Opt. Lett.,* 6, p. 294, 1981.

Smith, P.W., "On the physical limits of digital optical switching and logic elements," *BSTJ* 61, pp. 1975–1993, 1982.

Soffer, B.H. et al. "Associative holographic memory with feedback using phase-conjugate mirrors," *Opt. Lett.,* vol. 11, p. 118, 1986.

Steriti, R., Coleman, J. and Fiddy, M.A., "High resolution image reconstrution based on a fully connected architecture," *Inverse Problems,* vol. 6, p. 453, 1990.

Streibl, N. et al. "Digital optics," *Proc. IEEE,* 77, p. 1954, 1989.

Tompkin, W.R., Malcuit, M.S. and Boyd, R.W. "Enhancement of the nonlinear optical properties of fluorescin doped boric-acid glass through cooling," *App. Opt.,* 29, p. 3921, 1990.

Wherrett, B.S., and Tooley, F.A.P. *Optical Computing.* Proc. 34th Scottish Universities Summer School in Physics, Edinburgh University Press: Edinburgh, 1989.

Yeh, P. et al. "Photorefractive nonlinear optics and optical computing," *Opt. Eng.,* vol. 28, p. 328, 1989.

Zhang, H.-J. et al. "Self-focusing and self-trapping in new types of Kerr media with large nonlinearities," *Opt. Lett.,* vol. 14, p. 695, 1989.

Multidimensional Systems and Signal Processing, 2, 373–378 (1991)
© 1991 Kluwer Academic Publishers, Boston. Manufactured in The Netherlands.

Space-Time Complexity in Optical Computing*

H.J. CAULFIELD
Center for Applied Optics, University of Alabama in Huntsville, Huntsville, Alabama 35899

Received August 13, 1990, Accepted December 10, 1990

Abstract. Given certain simple and well defined operations and complexity measures, the product of spatial complexity with temporal complexity must exceed a certain minimum problem complexity if that processor is to solve that problem. Some optical processors violate that condition in a favorable direction (anomalously small temporal complexity). We next extend the requirement to embrace those optical processors. In its final form, the theorem requires that the product of spatial, temporal, and fanin complexities equal or exceed the problem complexity.

1. Introduction

We seek here to explore in detail the tradeoff between spatial and temporal complexity in optical computing. After defining some terms, we derive what appears to be a fundamental relationship between the two types of complexity. We then show how optical computers can perform better than this theoretical optimum. This, in turn, provides one more clear advantage of optics over electronics. Like the fanin and fanout advantage [Caulfield 1987] and the energy advantage [Caulfield and Shamir 1989], this advantage stems from the non-interfering nature of the optical beams.

2. Defining the Complexity Terms

There are no uniform definitions of the various type of complexity we wish to discuss. The definitions offered here are reasonable and consistent with many other definitions. Complexity has to do with resource allocation. The resources are assumed to occur in units (of time, space, etc.). We normalize these unit resources, so it is only the number of units we count. Thus the complexity of resource x is called χ_x and is the number of units of x required.

We will now do one more normalization. All complexities will be measured in terms of the input complexity χ_I. Let the smallest number governing the scaling of the input be N. For a linear array, $\chi_I = N$. For a rectangular array, $\chi_I = N^2$, etc.

In general, for a given N,

$$\chi_x = f_x(N). \tag{1}$$

*Some parts of this work were presented at the SPIE symposium "Optical Information Processing Systems and Architectures II," in San Diego, California, July 1990.

Let the asymptotic limit of $f_X(N)$ be called $O_X(N)$. We will call the x complexity

$$C_X = O_X(N). \tag{2}$$

It shows how the resource x scales with the input measure N for large N—the only case of interest in optical computing. We now look at some particular xs.

The area comlexity enters in when multiple operators exist side-by-side: the usual case for parallel optical computing. If the area per detector is one, we can have a normalized area

$$A = N^2. \tag{3}$$

This is, of course, the maximum spatial complexity in two dimensions, i.e.,

$$\chi_A \leq A = N^2 \tag{4}$$

and

$$C_A \leq O(N^2). \tag{5}$$

Thus we sometimes have unit $[O(1)]$ or linear $[O(N)]$ optical systems as well as the more common $O(N^2)$ systems.

The temporal operations in an optical processor may occur rapidly (10^{-12} sec) or slowly (10^{-2} sec). Let us normalize the time T required in terms of those unit steps. That is

$$\chi_T = T \tag{6}$$

where T is the number of time steps required. A pipelined operation may be $O(1)$ in time. A bit serial processor may be $O(N^3)$ or worse.

For the processor and algorithm chosen, there is a minimum number of operations required. We call this the processor-algorithm complexity χ_{PA}.

We know there must be a problem complexity

$$\chi_P = \min \{\chi_{PA}\} \tag{7}$$

over all, possible processor-algorithm pairs. Naturally, determining χ_P explicitly will be hard—perhaps impossible. But we do know that such a number exists.

3. Tradeoff Constraints

For any processor which solves a problem P, it seems obvious to require

$$\chi_A \chi_T \overline{>} \chi_f. \tag{8}$$

Solving P requires at least χ_P steps, by definition. The most steps that can be done in parallel is χ_A. Therefore χ_T, the number of time steps, must satisfy

$$\chi_T \geq \chi_P/\chi_A. \tag{9}$$

This observation somewhat related to various VLSI design rules on A, AT, and AT^2 which we will not discuss here, because the emphasis there is enough different from ours as to cause confusion. We mention this not to add clarity but to prevent confusion. A nice review of A, AT, and AT^2 rules and their logic has been given by Ullman [1984].

For a Turing Machine, $\chi_A = 1$ and we require $\chi_T \geq x_P$. For the best possible parallel processors, $\chi_A = N^2$ and $\chi_T \geq \chi_P/N^2$. Since optics deals with large N (10^2 to 10^3), the N^2 factor can be huge.

If the asymptotic dependencies are $C_A = O(N^\alpha)$, $C_T = O(N^\tau)$, and $C_P = O(N^\pi)$; we require

$$\alpha + \tau \geq p \tag{10}$$

For example, solving the set of linear equations

$$A\mathbf{x} = \mathbf{b} \tag{11}$$

for the unknown vector $\mathbf{x}$ is widely believed to satisfy $C_P = O(N^3)$. A fully parallel solver would give $C_A = O(N^2)$. It follows that the time required must satisfy

$$C_T = O(N^\tau) \tag{12}$$

where $\tau \geq N$.

4. Refuting the Tradeoff Constraint Analysis Just Offered

The analysis we offered can be rephrased as follows. χ_P steps are necessary. Those that we do not do in space, we must do in time. The logic seems unassailable.

We will show that this concept is wrong or, better, not applicable to some optical processors. To do this, we will offer one of many counter examples. We then show why optics can beat the tradeoff constraints of Expression 8.

The example of a system beating the tradeoff constraint theorem is the Bimodal Optical Computer or BOC [Caulfield, Gruninger, Ludman, Steiglits et al. 1986; Abushagur and Caulfield 1987; Abushagur, Caulfield, Gibson and Habli   ]. The system is sketched in Figure 1. There is an N component input vector connected to an N component output vector through an N^2 component (pixel) spatial light modulator. The optical matrix-vector finds

$$\mathbf{y} = A\mathbf{x} \tag{13}$$

quite crudely. The difference $\mathbf{b} - \mathbf{y}$ drives $\mathbf{x}$ to make $\| \mathbf{bv} - A\mathbf{x} \| = O$, also quite crudely. However, under circumstances too complex to describe here, we can cause this system to generate an $\mathbf{x}$ satisfying

$$\| \mathbf{b} - A\mathbf{x} \| < \epsilon, \tag{14}$$

21

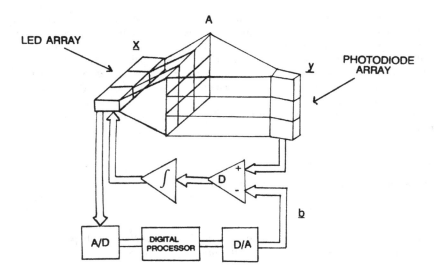

Figure 1. The Bimodal Optical computer.

for any preselected ϵ. The keys are to use a deliberately noisy A in the optical portion of the process and to use digitally assisted iterative improvements. The time taken between the insertion of input data and the extraction of a suitable output x is small but problem dependent. For singular or near-singular matrices, it is technically chaotic. That is, the output vector will be satisfactory, but which of the infinity of satisfactory vectors is produced and how long it takes to produce it is not predeterminable. What we do know, however, is that that time is N-independent, i.e.,

$$C_T = O(1) = O(N^0). \tag{15}$$

Of course,

$$C_A = O(N^2). \tag{16}$$

Then

$$\alpha - T = 2 + 0 = 2, \tag{17}$$

while the problem complexity is roughly $O(N^3)$,[1] so

$$\alpha + \tau(=2) < \pi(\sim 3). \tag{18}$$

The tradeoff constraint of expression 8 is violated.

Violating anything so fundamental and obvious as that tradeoff constraint must mean that something fundamental is going on which was ignored during the earlier analysis. That is, indeed, the case.

The simple argument of Section 3 assumed that the A and T complexities could be accounted for separately. In massively interconnected optical systems, the complexity of the operation occurring in any unit time increases with A. That is, as A increases, the number of data processed at each output increases.

In the case of the BOC, the complexity associated with each operation is proportional to N. That is, the fanin is N. The new complexity tradeoff is

$$\chi_A \chi_P \chi_F \geq \mathfrak{x}_P. \tag{19}$$

Thus if

$$C_F = O(N^\phi), \tag{20}$$

we also have

$$\alpha + \tau + \phi \geq \pi \tag{21}$$

In the BOC case $\alpha = 2$, $\tau = 0$, $\phi = 1$, and $\pi = 3$. Thus

$$\alpha + \tau + \phi = 3 \tag{22}$$

as required.

In fully connected optical systems, $\phi = 2$ [Caulfield 1987; Burden and Faires 1989]. Then, if we can use all of this parallelism and fanin effectively, we have

$$\chi_T \leq \chi_P / N^4 \tag{23}$$

or

$$\tau \leq \pi - 4. \tag{24}$$

Most problems which are not exponential have $\pi \leq 4$. That is, we cannot exclude in principle solving $O_P(N^4)$ problems $O_T(1)$. As yet, no good example of such a solution is known to us.

5. Conclusions

For most processors, the product of the time and area complexities must equal or exceed the problem complexity to achieve success. Because fanin in optics is essentially unlimited (actually practically limited to a little over 10^{12} as shown in [Shamir, Caulfield and Johnson 1989]) and because the complexity of the operation performed scales with the fanin, we are driven to a new relationship which requires the product of fanin complexity, area complexity, and time complexity to equal or exceed the problem complexity.

This constitutes a new unique advantage to optics over electronics. The prior advantages being the size of the fanin available [Caulfield 1987; Shamir, Caulfield and Johnson 1989] and the consequent decrease in energy requirements for some given computations [Caulfield and Shamir 1989].

Notes

1. The most efficient method for solving $Ax = b$ known to me is the Cholesky method which has a number of operations whose leading term is $N^3/6$.

References

M.A.G. Abushagur, H.J. Caulfield, D.M. Gibson, and M. Habli, "Superconvergence of hybrid optoelectronic processors."

M.A.G. Abushagur and H.J. Caulfield, "Speed and convergence of bimodal optical computers," *Opt. Eng.* 26, pp. 22–27, 1987.

R.L. Burden and J.D. Faires, *Numerical Analysis*, 4th Edition, PWS-Kent: Boston, pp. 370–372, 1989.

H.J. Caulfield and J. Shamir, "Wave particle duality considerations in optical computing," *Appl. Opt.*, vol. 28, no. 12, pp. 2184–2186, 1989.

H.J. Caulfield, "Parallel N^4 weighted optical interconnections," *Appl. Opt.* 26, p. 4039, 1987.

H.J. Caulfield, J.H. Gruninger, J.E. Ludman, K. Steiglits, H. Rabitz, J. Gelfand, and E. Tsoni, "Bimodal optical computers," *Appl. Opt.* 25, pp. 3128–3131, 1986.

J. Shamir, H.J. Caulfield and R.B. Johnson, "Massive holographic interconnection networks and their limitations," *Appl. Opt.* 28, pp. 311–324, 1989.

Jeffrey D. Ullman, *Computational Aspects of VLSI*, Computer Science Press: Rockville, Maryland, 1984.

Multidimensional Systems and Signal Processing, 2, 379–390 (1991)
© 1991 Kluwer Academic Publishers, Boston. Manufactured in The Netherlands.

Composite Reference Image for Joint Transform Correlator

DAVID MENDLOVIC, NAIM KONFORTI AND EMANUEL MAROM
Faculty of Engineering, Tel Aviv University, Tel Aviv, Israel, 69978

Received October 10, 1990, Accepted February 11, 1991

Abstract. Extension of the joint transform correlator (JTC) operation to include a complex reference image has been presented. Such amplitude and phase reference images are common when using a single harmonic (circular, radial, logarithmic, etc.) in correlation setups or when implementing composite filters. The analysis of an aspect view (tilt) invariant pattern recognition system, using logarithmic harmonics decomposition, JTC principles and composite filter techniques, is described followed by experimental results that can be obtained in real time.

1. Introduction

The optical joint transform correlator (JTC) was proposed several years ago by Weaver and Goodman [1966]. In that correlator the reference image as well as the unknown object are presented simultaneously in the input plane, and their joint transform is produced in the focal plane behind a transform lens. The joint transform is recorded on a square law detector (originally a photographic film) and using a simple Fourier transform configuration, behind another transform lens, the correlation plane is produced in the first diffraction order. The main advantage of such system is that one does not need a matched filter, one simply introduces the reference image, in the common case a real function, side by side with the input object in the input plane. A disadvantage of the JTC is that the space bandwidth product available in the input plane must be shared by the input object, the reference image as well as by a safety band to ensure separation of the correlation pattern from undesired terms at the output plane.

In recent years, the use of joint transform correlators has become popular, in particular, due to improvements in spatial light modulator (SLM) devices. Yu and Liu [1984] applied this technique to real time pattern recognition configurations by using SLMs in optical systems.

Joint transform correlators preserve the shift invariant pattern recognition property, but have high sensitivity to other invariant parameters such as rotation, scale or projection. It has been shown [Hsu and Arsenault 1982; Mendlovic, Marom and Konforti 1988; Mendlovic, Konforti and Marom forthcoming] that the classical correlator configuration can be made invariant to one additional parameter by using a matched filter containing only a part of the input data. This added parameter could be the angular orientation of an object when the correlation is performed with respect to only one harmonic out of the circular harmonic decomposition [Hsu and Arsenault 1982], or the scale of the object when the matched filter consists of only one harmonic out of the radial harmonic decomposition

25

[Mendlovic, Marom and Konforti 1988]. Recently we described two other types of harmonic decompositions, the logarithmic harmonic [Mendlovic, Konforti and Marom, forthcoming] which enables projection (i.e., aspect view) invariant pattern recognition and the more general deformation harmonic [Marom, Mendlovic and Konforti, forthcoming] that can handle any type of distortion parameters (including those mentioned above). We also showed [Mendlovic, Marom and Konforti, forthcoming] that all these harmonic decompositions can be used in joint transform correlator configurations.

In conventional correlator configurations the use of a composite filter [Caulfield and Maloney 1969], or in a more general sense, a synthetic discriminant functions (SDF) [Kumar 1988], permits the recognition of objects from one classification while ignoring objects from another. These methods have also been successfully implemented for rotation invariant pattern recognition [Arsenault and Hsu 1983] by using a composite circular harmonic filter.

Recently, Javidi [1989] has shown a bipolar nonlinear joint transform correlator that permits the use of a SDF as a reference object. His SDF based binary JTC uses thresholding at both the input and the Fourier planes, but requires a real reference object. However, the most general reference object has phase information, as for instance, when it describes a composite image (derived in a similar way as the composite filter was defined), or when the reference object contains one harmonic (circular, radial or logarithmic) generated from the desired pattern.

We now describe the composite reference image approach for use with joint transform correlators. This approach requires the calculation of a synthetic composite image to be used as a reference object when placed side by side with the input object in a common plane. In addition we describe the using of reference composite harmonics for the case of logarithmic harmonic functions. Since the composite image as well as the composite harmonics contain, in general, phase as well as amplitude data, the conventional joint transform correlator configuration cannot be used, and new configurations should be conceived.

In a previous paper [Mendlovic, forthcoming] we presented an elegant way that allows the use of joint transform correlators for rotation invariant pattern recognition. There, we used as reference object a single harmonic out of the circular harmonic decomposition of the desired object, and showed how recording of phase as well as amplitude information is achieved by encoding them on a grating (via Lohmann's phase detour model [Lohmann and Paris 1967]). The input object, placed side by side with this reference, has to be multiplied by a grating too, so that the joint transform correlation distribution will now be obtained at the location of the first diffraction order.

One should note that here we encode, using computer generated holograms (CGH), the harmonic itself and not its spectra as in the more conventional matched filter.

In this paper we deal with two cases of recognizing objects from one group and rejecting objects from another one. The first example consists of a straight composite image case when we try to recognize the letter E and ignore the letter P. The second example deals with the same letters but those can now be tilted along one axis. This case can be defined as aspect view invariant joint transform correlators, and requires using composite logarithmic harmonics. Section 2 presents and justifies mathematically the optical setup, Section 3 briefly describes the logarithmic harmonics functions, Section 4 presents the composite image and the composite logarithmic harmonics approach, and in Section 5 some optical experimental results are presented and discussed.

2. The Optical Setup

A computer generated hologram (CGH) that carries the reference image is placed side by side with the input pattern in a common input plane. The basic nature of the CGH is that they display the encoded information along a diffraction order in an off-axis direction and thus the input pattern has to be multiplied by a grating in order to generate the spectrum along the same diffraction direction. If the input pattern is real, as in our case, a simple multiplication by a grating is sufficient. Complex input patterns would need to be encoded holographically in any event, and thus another CGH would be constructed for them too. The two diffracted beams propagating along parallel paths are those that will generate the joint transform correlation. An experimental setup able to perform this operation in real time is sketched in Figure 1.

The input plane contains the input pattern multiplied by a uniform grating, and along one of its sides, a specific CGH representing the reference image. As a result, the amplitude transmittance of the input plane can be expressed as

$$t(x, y) = [f_{in}(x, y - y_0/2) + f_r(x, y + y_0/2)](1 + \cos (2\pi\alpha_0 x)) \tag{1}$$

where $f_{in}(x, y)$ is the pattern that has to be recognized, $f_r(x, y)$ is the reference image, y_0 is the distance between the center of the input pattern and the reference harmonic center, and α_0 is the grating frequency. Although the grating is explicitly written outside the common parentheses in Equation 1, in most cases it is embedded in the f_r expression where it is used to encode phase as well as amplitude information. The intensity distribution in the Fourier transform plane, along the first diffraction order in the x direction is

$$I(\alpha, \beta) = |F_{in}(\alpha - \alpha_0, \beta)e^{-(i2\pi/\lambda f)(y_0/2)\beta} + F_r(\alpha - \alpha_0, \beta)e^{(i2\pi/\lambda f)(y_0/2)\beta}|^2$$

$$= |F_{in}(\alpha - \alpha_0, \beta)|^2 + |F_r(\alpha - \alpha_0, \beta)|^2$$

$$+ F_{in}(\alpha - \alpha_0, \beta)F_r^*(\alpha - \alpha_0, \beta)e^{-(i2\pi/\lambda f)y_0\beta}$$

$$+ F_{in}^*(\alpha - \alpha_0, \beta)F_r(\alpha - \alpha_0, \beta)e^{(i2\pi/\lambda f)y_0\beta} \tag{2}$$

where F_{in} and F_r are the Fourier transforms of f_{in} and f_r respectively, (α, β) are the spatial coordinates in the Fourier plane, λ is the illumination wavelength and f is the lens focal

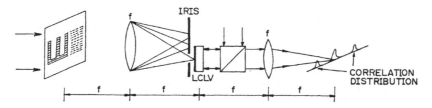

Figure 1. Set-up for a real-time joint transform correlator.

length. The Fourier transform of the last two terms in Equation 2 exhibits the desired correlation of the input f_{in} and the reference image f_r at the location of the first diffraction orders in the y direction (see Figure 1). In our configuration, a liquid crystal light valve (LCLV) is positioned in the first Fourier transform plane and the joint transform is projected on its photoconductive layer (input side). The read out beam of the LCLV is proportional to the intensity (Equation 2) of the input side, thus generating the correlation distribution of the two functions in real time (Figure 1) by virtue of the Fourier transform of the last two terms of Equation 2.

We chose to handle the aspect view (tilt) invariance of this work and thus decomposed an object $f(x, y)$ into logarithmic harmonics terms [Mendlovic, Konforti and Marom, forthcoming] as shown in the next section.

3. Logarithmic Harmonic Functions

Following the derivation presented earlier by us [Mendlovic, Konforti and Marom, forthcoming], we suggest to decompose an object into the following logarithmic harmonic set

$$\phi_N(x, y) = f_N(y)x^{i2\pi N - 1/2} = f_N(y)e^{(i2\pi N - 1/2)\ln(x)} \tag{3}$$

when $N = 0, \pm 1, \pm 2 \dots$.

The one dimension projection invariant procedure should be used whenever the object is expected to be tilted around a single axis. For simplicity, let use assume that this axis is the principal axis y. In order to accommodate nonpositive values of x, we have to define a modified logarithmic harmonic as indicated below

$$\phi_N(x, y) = f_N(y)|x|^{i2\pi N - 1/2} = f_N(y)e^{(i2\pi N - 1/2)\ln(|x|)} \tag{4}$$

We will now normalize the x coordinate using the following limits: $x = 1$ is the size of the pattern at its original maximal scale and $x = e^{-L}$ is the smallest reducing factor, where L is a positive integer. It is easy to prove that this set of logarithmic harmonics is orthogonal and complete [Mendlovic, Konforti and Marom, forthcoming].

Using the expansion set of Equation 4, a two dimensional object $f(x, y)$ can be decomposed into

$$f(x, y) = \sum_{N=-\infty}^{\infty} \phi_N(x, y) = \sum_{N=-\infty}^{\infty} f_N(y)|x|^{i2\pi N - 1/2} \tag{5}$$

with

$$f_N(y) = \int_{-1}^{e^{-L}} f(x, y)(-x)^{-i2\pi N - 1/2} \, dx + \int_{e^{-L}}^{1} f(x, y)x^{-i2\pi N - 1/2} \, dx \tag{6}$$

The decomposition is done for objects tilted around the y axis, thus having scaled values along the x direction. When a single logarithmic harmonic is selected and used as a matched filter, the correlation plane is invariant to aspect view changes of the input plane. (For more detailed explanation see [Mendlovic, Konforti and Marom, forthcoming]).

4. Composite Image and Composite Harmonic

A composite filter [Caulfield and Maloney 1969] is essentially a linear combination of filters associated with several objects. Such a filter can be used to improve the performance of a conventional matched filter, as for instance for recognizing an object while rejecting others. We suggest to use this composite filter concept in the image plane and thus called it *composite image*. A composite image can also be generated when using their logarithmic harmonics, provided that we choose the same harmonic for all patterns. For two components, a composite logarithmic harmonic image is:

$$f_r(x, y) = [af_N(x, y) + bg_N(x, y)] |x|^{i2\pi N - 1/2} \tag{7}$$

where a and b are complex constants found by solving the following matrix equation:

$$\begin{bmatrix} C_{ffN} & C_{fgN} \\ C_{gfN} & C_{ggN} \end{bmatrix} \begin{bmatrix} a \\ b \end{bmatrix} = \begin{bmatrix} R_1 \\ R_2 \end{bmatrix} \tag{8}$$

R_1 and R_2 are the outputs required when the inputs are $f(x, y)$ and $g(x, y)$ respectively and C_{ffN}, C_{ggN}, C_{fgN}, C_{gfN} are the central $(0, 0)$ value of the crosscorrelation between f and g and their Nth logarithmic harmonic respectively. In most practical cases R_1 and R_2 are 1 and 0 respectively.

One should note that a composite logarithmic harmonic may contain more than two images, but to conserve the aspect view invariance property, all the terms must be related to the same harmonic order.

5. Experimental Results

5.1. Composite Image Joint Transform Correlation

Utilizing the optical setup presented in Figure 1 we have shown that it can be used for recognizing one object—the letter E, and rejecting another object—the letter P. The reference composite image was computed using these two letters E and P with different weights (0.867 for the E and -0.677 for the P) and encoded as a CGH using Lohmann's detour phase method [Lohmann and Paris 1967] with a resolution of 64×64 pixels and 30 levels of amplitude and phase. The input object was multiplied by a grating with the same spatial frequency as the carrier of the CGH and placed side by side with the CGH. The reference image was plotted using a standard laser printer and then photoreduced to a size of 10×5 mm.

Figure 2a shows the view of the input plane where a composite image was used as a reference object side by side with the input object—the letter E. Figure 2b shows the correlation results. In the upper part of the figure, the profile of the correlation plane is displayed. Figure 3 shows the same when we input the letter P but this time we got no significant peak in the correlation plane. All records were performed with the same CCD camera setting.

For comparison, we repeated the experiment with a reference object containing just the letter E. Figure 4 shows the correlation case and Figure 5 represents the crosscorrelation case. This time one notes that bright correlation peak appears also for the letter P.

5.2. Aspect View Invariant Joint Transform Correlation

To demonstrate the aspect view invariant JTC with improved selectivity, two cases were tested, the first designed for the recognition of the letter E in various tilted versions, and the second for the recognition of the letter E and rejection of the letter P also in various 1-D distorted versions. The reference image in the JTC setup consisted, in the first case, of a single logarithmic harmonic of the letter E and in the second case of the composite image of two logarithmic harmonics of same order with different weights, one derived for the letter E and the second for the letter P. In both cases the logarithmic harmonics were encoded in a CGH form with the same resolution as before.

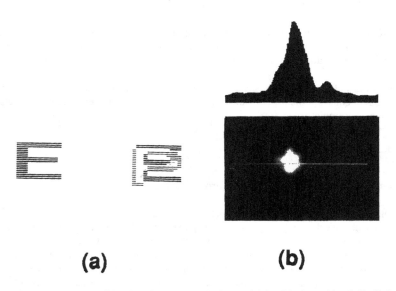

(a)   **(b)**

Figure 2. (a) Input plane containing the reference composite image (right) and the input object (left). (b) Correlation result.

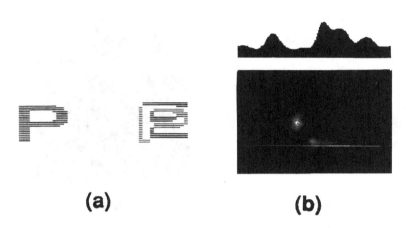

(a)   **(b)**

Figure 3. Same as Figure 2 but with the letter *P* as input.

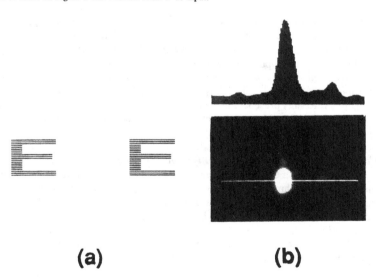

(a)   **(b)**

Figure 4. Same as Figure 2 but without using the composite reference object method.

5.2.1. *Logarithmic Harmonic Reference.* We will now present experimental results obtained with a JTC setup and a single logarithmic harmonic of the letter *E*. Aspect view invariant pattern recognition was achieved but the system has difficulties in rejecting close patterns,

31

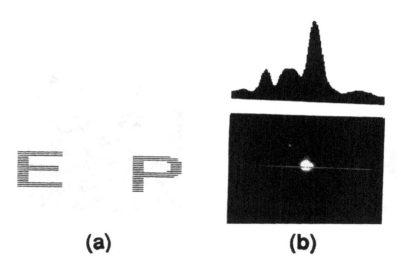

Figure 5. Same as Figure 4 but with the letter *P* as input.

due to high crosscorrelation with similar letters. Figure 6 shows the different input patterns that were used in these experiments. Figures 6a–c are different tilted versions of the letter *E* with aspect ratio of 1.0, 1.25, and 0.5 respectively and Figure 6d is the letter *P*. Figure 7 presents the view of the input plane where a single logarithmic harmonic was used as a reference object side by side with the letter *E*. Figure 8 displays various correlation and crosscorrelation results when Figure 6 was used as input (placed at the location of the letter *E* in Figure 7). Results presented in Figure 8 prove the logarithmic harmonic method can be incorporated in JTC with good aspect view invariant capabilities but with poor rejection of patterns similar to the stored one.

5.2.2. Composite Logarithmic Harmonic Reference. To improve the selectivity of the system, we suggested using a composite reference image that contains a linear combination of the same order logarithmic harmonics of the letters *E* and *P*. Figure 9 shows the input plane that contains the composite image as well as the tested object.[1] Figure 10 shows the crosscorrelation results when the inputs of Figure 6 were used. The correlation results display a bright correlation peak in Figures 10a–c, while the crosscorrelation result of Figure 10d displays no peak value, as desired, thus accomplishing the recognition of the letter *E* in all its tilted versions, while rejecting the letter *P*.

6. Conclusions

The real-time joint transform correlator has been extended to include complex reference images. We also incorporated the composite filter method using the composite image reference object so that the selectivity is improved. Applying this concept with logarithmic

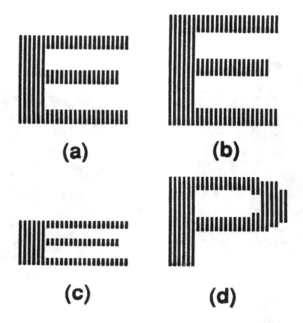

Figure 6. Patterns used for recognition tests: (a), (b) and (c) are different tilted versions of the letter *E* with the aspect ratio (a) 1.0 (b) 1.25 and (c) 0.5. (d) is a pattern for crosscorrelation test.

Figure 7. Input plane containing the selected logarithmic harmonic (right) and the input object (left.)

harmonic functions, aspect view invariant pattern recognition is achieved. These approaches were successfully tested by encoding the reference image and the input objects as computer generated holograms, which indeed require systems with larger space bandwidth capabilities, but present day technology is able to provide it.

Acknowledgment

The authors wish to thank Mr. M. Deutsch for his technical support of part of this work. The authors thankfully acknowledge partial support received from the Barry Zukerman Research Fund established by Symbol Technologies, Inc., Bohemia, NY, as well as from a grant from the European Community.

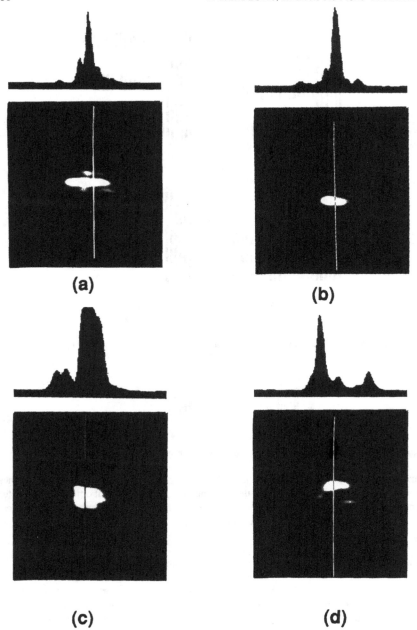

Figure 8. Crosscorrelation results using the input layout of Figure 7, but with the input objects shown in Figure 6.

Figure 9. Same as Figure 7 but with a composite harmonic reference object.

Notes

1. The linear complex weights were $(-0.239 + i0.063)$ for the E's second logarithmic harmonic and $(0.115 + i0.068)$ for the P's second logarithmic harmonic.

References

H.H. Arsenault and Y.N. Hsu, "Rotation invariant discrimination between almost similar objects," *Appl. Opt.* 22, pp. 130–132, 1983.

H.J. Caulfield and W.T. Maloney, "Improved discrimination in optical character recognition," *Appl. Opt.* 8, pp. 2354–2356, 1969.

Y.N. Hsu and H.H. Arsenault, "Optical pattern recognition using circular harmonics expansion," *Appl. Opt.* 21, pp. 4016–4019, 1982.

B. Javidi, "Synthetic discriminant function-based binary nonlinear optical correlator," *App. Opt.* 28, pp. 2490–2493, 1989.

B.V.K.V. Kumar, "Review of synthetic discriminant functions," *Proc. Soc. Photo-Opt. Instrum. Eng.* 960, pp. 18–28, 1988.

A.W. Lohmann and D.P. Paris, "Binary Fraunhofer holograms," *Appl. Opt.* 6, pp. 1739–1748, 1967.

E. Marom, D. Mendlovic and N. Konforti, "Generalized spatial deformation harmonic filter for distortion invariant pattern recognition," accepted for publication in *Opt. Commun.*

D. Mendlovic, E. Marom and N. Konforti, "Shift and scale invariant pattern recognition using Mellin radial harmonics," *Opt. Commun.* 67, pp. 172–176, 1988.

D. Mendlovic, N. Konforti and E. Marom, "Shift and projection invariant pattern recognition using logarithmic harmonics," accepted for publication in *Appl. Opt.*

D. Mendlovic, E. Marom and N. Konforti, "Invariant joint transform correlator," accepted for publication in *Opt. Lett.*

C.S. Weaver and J.W. Goodman, "A technique for optically convolving two functions," *App. Opt.* 5, pp. 1248–1249, 1966.

F.T.S. Yu and X.J. Liu, "A real-time programmable joint transform correlator," *Opt. Commun.* 52, pp. 10–16, 1984.

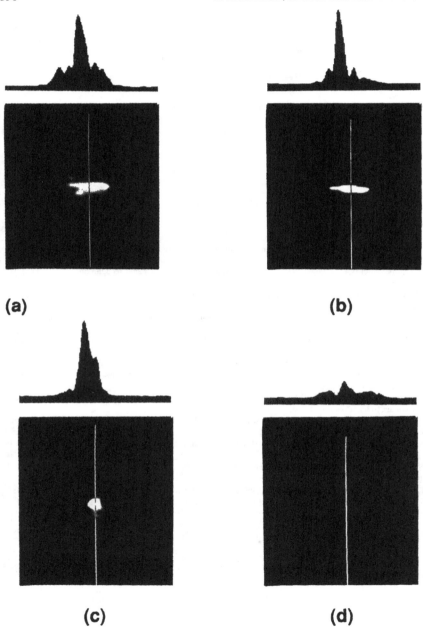

Figure 10. Same as Figure 8 but with Figure 9 as the input layout.

Multidimensional Systems and Signal Processing, 2, 391–400 (1991)
© 1991 Kluwer Academic Publishers, Boston. Manufactured in The Netherlands.

Optical Computation of the Sector and Radon Transform Using Pinhole Array

DEJAN V. PANTELIC

Institute of Physics, Maksima Gorkog 116, Zemun, P.O.B. 57, 11001 Belgrade, Yugoslavia

Received October 30, 1990, Accepted February 11, 1991

Abstract. An optical setup for the computation of the sector transform is presented in this paper. The device is based on pinhole imaging and processing with the computer generated mask. Computation of the transform does not require image rotation, and therefore processing is done fully in parallel. Extensions of the transform are also discussed.

1. Introduction

Some modifications of tomographic algorithms have been used in image processing [Eichmann 1987]. Recently, we have shown that the sectors of an image offer interesting possibilities in noise-insensitive pattern recognition [Pantelic 1988]. The proposed transform is a natural extension of the Radon transform, and there is a one-to-one relationship between them.

Radon and sector transforms are time consuming, if computed on a classical computer. Therefore, we previously proposed an optical setup for the computation of the sector transform [Pantelic 1989]. The optical system was constructed by using anamorphic optics and a triangular mask. As in the case of the optical Radon transform computing [Easton, Ticknor and Barrett 1984], the input image had to be rotated, consequently decreasing the computational speed.

In this paper we will present an optical setup, capable of computing the sector transform, fully in parallel.

2. The Theory of the Sector Transform

In this section we will make a definition of the sector transform and, also, show that it is noise insensitive (to a certain extent). But first, we will explain the motivation standing behind the sector transform.

Figure 1a shows a binary image. Assume that this original pattern is corrupted by noise (Figure 1b). We will search for some features of this image which are not greatly influenced by the noise. The length of the perimeter of the image is certainly *not* a feature, since it can be a fractal curve (thus with infinite length). But the surface of the image can be expected to remain almost the same. This is because some parts of the original image are

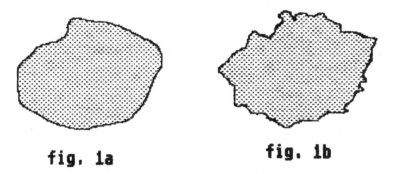

Figure 1.

missing, however new parts are added to the image. In the statistical sense, area should remain the same. Of course, these notions were made on an intuitive level. We shall try to prove these facts in a more formal way.

The Hough transform is defined by the length (L) of the intersection between the image and a straight line (Figure 2a), when the line is changing its position (r) and orientation (θ). Similarly, the sector transform is defined as an area (S) of the image, after intersecting with the line 1, for various positions and orientations (Figure 2b).

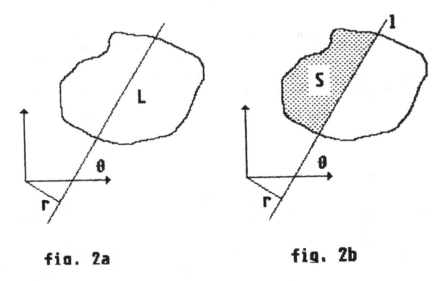

Figure 2.

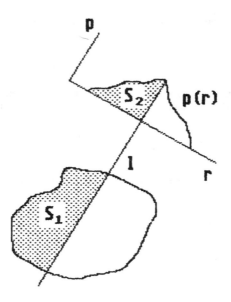

fig. 3

Figure 3.

The sector and the Hough transforms are connected. We shall look at one Hough projection and line 1 in this projection (Figure 3). It is easy to see that areas S_1 and S_2 are equal. Consequently, one sector projection is obtained by integrating on Hough projection:

$$S(r) = \int_o^r p(r)dr \tag{1}$$

It will be studied how the sector transform behaves in the presence of the noise. We will not study the image itself, but its Hough projection and, further, sector projection, using formula (1). It will be assumed that the noise in the image is transferred to its Hough projection. Several assumptions about the noise will be made: that it is an additive, stationary stochastic process with zero mean.

In other words, if the Hough projection is given by the function $p(r)$, then its noisy version is:

$$p_1(r) = p(r) + n(r) \tag{2}$$

In the first step, we shall find the sector transform of the noise $n(r)$:

$$S(r) = \int_0^r n(r)dr \tag{3}$$

since $n(r)$ is zero mean stochastic process, it is easy to prove that the mean value of $S(r)$ is also zero [Papoulis 1965]. The standard deviation of $S(r)$ is given by [Papoulis 1965]:

$$\sigma^2 = \int_0^r \int_0^r (R(r_1, r_2) - e(r_1) e(r_2))dr_1 \, dr_2 \tag{4}$$

where $e(r)$ is mean value of $S(r)$. Since $e(r) = 0$ and, by the definition of a stationary stochastic process:

$$R(r_1, r_2) = R(r_1 - r_2) \tag{5}$$

then:

$$\sigma^2 = \int_0^r \int_0^r R(r_1 - r_2)dr_1 \, dr_2 \tag{6}$$

There is a well known relationship between the correlation R and correlation coefficient:

$$R(r_1, r_2) = \sigma_0^2 K(r_1, r_2) \tag{7}$$

where σ_0 is a standard deviation of a noise $n(r)$.
Therefore:

$$\sigma^2 = \sigma_0^2 \int_0^r \int_0^r K(r_1 - r_2)dr_1 \, dr_2 \tag{8}$$

If the process is uncorrelated, i.e.:

$$K(r_1 - r_2) = \begin{array}{l} 1 \text{ if } r_1 = r_2 \\ 0 \text{ if } r_1 \neq r_2 \end{array} \tag{9}$$

then $\sigma^2 = 0$, which means that there is no noise in the sector transform (regardless of the noise in the original image). Of course, assumption (9) is not realistic and therefore, correlation coefficient should be represented by a certain distribution. If the distribution is narrow (i.e., random process is weakly correlated) then the integral in (8) is small, thus diminishing the noise. Otherwise, if distribution is wide (the noise is highly correlated) noise can be amplified. However, it can be shown [Pantelic 1988] that, even in that case, signal-to-noise ratio is improved.

3. Sector Transform Computation Using Pinhole Array

A pinhole array has been successfully used in various optical processing tasks [Lee 1981]. The principle is simple: image is multiplexed by using the pinhole array, and processing is done individually on each copy of the input image.

On the other hand, the sector transform is obtained by masking various portions of an input image (Figure 4) and calculating the area of unmasked portion. The mask edge is defined by the straight line 1 whose position is defined in the polar coordinate system (r, θ).

If we use the pinhole array for the sector transform calculation, the image processing step is rather simple: each copy of an input image is masked, as shown in Figure 5. However, the position of the line 1 should be moved in order to scan the whole (r, θ) plane, in a discrete number of points. Therefore, a replicated image should be masked by the composite mask—each submask corresponds to one discrete position of the line 1.

The required mask had to be computer generated and matched to the number of pinholes and dimensions of replicated images. In our case, we used pinhole array which was made by photolithographic techniques in a thin metal foil. There were 10×10 holes with $200 \ \mu m$ radius each. The appropriate mask was computer generated accordingly (Figure 6), and reduced photographically.

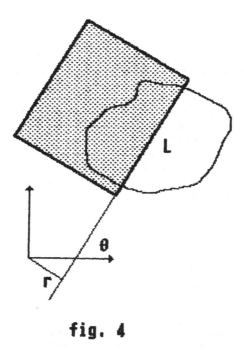

fig. 4

Figure 4.

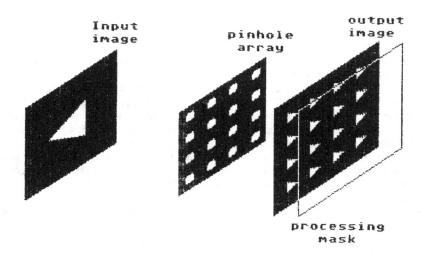

Figure 5.

As an example, we will present the experimental results of the optical sector transform of a triangular input image. In the output plane multiplicity of replicas was obtained (Figure 7a). After masking, the resulting image was recorded in the output plane of the processor (Figure 7b). In order to obtain the required sector transform, each of the subimages had to be integrated. This was done by using the photo-detector array, where the area of each detector is larger than any of the subimages.

The size of the individual images was matched to the size of detectors in array, by simply changing the distance between the pinhole mask and image plane.

In principle, the number of pinholes in array is arbitrary. It depends only on how finely we want to discretize the sector transform space. Of course, certain limitations exist: pinholes should not be placed too densely, since this will result in image overlapping; the size of the pinhole-array should not be much larger than the original image, since in that case replicated images would not be of the same intensity.

4. Discussion

In the previously described manner, we have succeeded in computing the sector transform in parallel, therefore significantly reducing the computational time. There are, however, certain sources of errors.

One error is inherent to pinhole imaging: each of the replicas of an input image is degraded by the point spread function (PSF) of the system. It is clear that the radius of PSF (which is circular in this case) must be smaller than the smallest feature in the image. This relationship, between the image and the PSF, can be controlled by changing the distance between the input plane, pinhole array and the output plane.

Figure 6.

Transforming the input image is just one step in a pattern recognition process. Another important step is feature space construction, which can be done in many ways. In our situation thresholding seems to be the most appropriate, since it can be done either optically or electronically. In this way we obtain the line which divides the sector transform into two areas: above and below the threshold. This results in a dimensionality reduction, necessary for the successful pattern recognition.

It should be noted that the sector transform can be further generalized by changing the type of the masking function. For example, line 1 could be curved (Figure 8a) or it could be replaced by the pair of lines resulting in a slit like function (Figure 8b). In the latter case, we can note that by letting $d \rightarrow 0$ we reduce the transform to the classical Radon transform. On the other end, by letting $d \rightarrow \infty$, we are approaching the sector transform. It is to be expected that, for intermediate values $0 < d < \infty$, resulting transformation

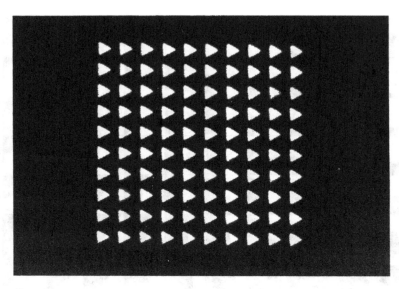

Figure 7a.

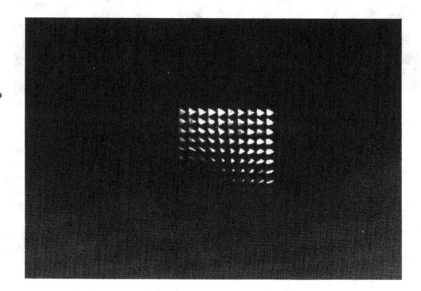

Figure 7b.

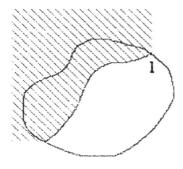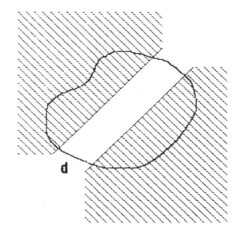

Figure 8.

should combine line detecting capability of Radon (Hough) transform and low noise sensitivity of the sector transform.

Another point also should be emphasized. Radon transform can be computed by simply finding the first derivative of the sector transform:

$$R(\theta,\, r) \;=\; -\,\frac{S(\theta,\, r)}{r} \tag{10}$$

It is more convenient to do so, due to much better light collection while computing the sector transform. When dealing with the pinhole array imaging, it is best to perform the first derivation on a digital computer. Due to the limited number of points this task will not require too much computer time.

5. Conclusion

We have shown that the sector transform can be performed optically, fully in parallel, by using pinhole imaging. Radon transform can be also computed, if simple postprocessing operations are performed.

References

R.L. Easton, A.J. Ticknor, and H.H. Barrett, "Application of the Radon transform to production of the Wigner distribution function," *Opt. Eng.* vol. 23, p. 738, 1984.

G. Eichmann, V. Li, "Real-time optical line detection," *Opt. Commun.* vol. 63, p. 230, 1987.

S.H. Lee, *Optical Information Processing.* Springer-Verlag: Berlin, 1981.

D.V. Pantelic, "Sector transform and its optical realization," *Opt. Commun.* vol. 74, p. 155, 1989.

D.V. Pantelic, "Sectors of a binary image used for noise insensitive pattern recognition," *Opt. Commun.* vol. 68, p. 257, 1988.

D.V. Pantelic, "Pattern recognition using statistical properties of sectors of an image," *Proc. SPIE*, (P.J.S. Hutzler, A.J. Oosterlinck, eds.) vol. 1027, p. 140, 1988.

A. Papoulis, *Probability, Random Variables and Stochastic Processes.* McGraw Hill: New York, 1965.

Multidimensional Systems and Signal Processing, 2, 401–419 (1991)
© 1991 Kluwer Academic Publishers, Boston. Manufactured in The Netherlands.

Optical Signal Processing Using Photorefractive Effect

KEN-ICHI KITAYAMA AND FUMIHIKO ITO
NTT Transmission Systems Laboratories, 1-2356 Take, Yokosuka-shi, 238-03 Japan

Received December 7, 1990, Accepted February 4, 1991

Abstract. Novel applications of photorefractive effect to optical signal processing are proposed and demonstrated. The applications of two-wave mixing (2-WM) in a bulk photorefractive crystal include logic operations and cross connect. Photorefractive crystal waveguide (PCW) is another primary concern. Holographic storage of Fourier transformed image in PCW and its application to optical neural network, two-dimensional array of PCW as a storage device, phase conjugate mirror using PCW, and structural consideration of PCW for efficient 2-WM are investigated. In the experiments, $LiNbO_3$ and $BaTiO_3$ waveguides are used. $BaTiO_3$ waveguide will be tested here for the first time. In a long-term prospect, all these technologies will eventually find important roles in optical signal processing.

1. Introduction

High spatial and temporal bandwidth of free-space optical interconnection as well as the potential of high-speed optical devices which electronics cannot match are a strong motivation to introduce the optical technologies to signal processing [Smith 1987]. Unique attributes of free-space optical interconnection including massive parallelism, broadcasting capability, and delay free, distortion free, and interference free characteristics [Goodman, Leonberger, Kung and Athale 1984] would be great benefits that can only be derived from the exploitation of optical technologies.

Optical parallel signal processing will become a promising candidate which meets the increasing needs of image processing having high processing capability [Arsenault, Szoplik and Macukow 1989] if the massive parallelism of optical interconnections is successfully introduced. In parallel processing, photorefractive effect could play the key role because it can serve not only for data storage material but also for optical interconnect device [Gunter and Huignard 1988; Fisher 1983].

Photorefractive crystal waveguide (PCW) is a novel approach to the enhancement of the nonlinear effects because the waveguide geometry provides unique advantages over bulk crystal due to a strong confinement of optical energy as well as long interaction length of waves [Hellwarth 1979; Hesselink 1990]. The waveguide geometry also allows one to synthesize compact devices into any configuration such as a matrix array and bundle [Hesselink and Redfield 1988]. Thus, the fabrication difficulty of a large single crystal still existing in some special class of crystals, can be circumvented. *SBN* [Hesselink and Redfield 1988] and $LiNbO_3$ [Yoshinaga, Kitayama and Oguri 1990] fibers have so far been successfully fabricated.

In this paper, novel applications of photorefractive effect, in the fields of optical signal processing which so far have not been explored, are proposed and demonstrated. First, the applications of two-wave mixing (2-WM) including logic operations and cross connect are focused on. Next, a photorefractive crystal waveguide (PCW) is investigated. The wave-guiding characteristics of PCW, holography of Fourier transformed image in PCW, and two-dimensional (2-D) array of PCW as a storage device and its application to optical neural network, and phase conjugate mirror using PCW, are presented. Finally, the structural consideration of PCW for efficient 2-WM is discussed. In the experiments, $LiNbO_3$ and $BaTiO_3$ waveguides are used. $BaTiO_3$, cleaved out of the bulk crystal, will be tested here for the first time. In a long-term prospect, all these technologies will eventually find important roles in optical signal processing.

2. Two-Wave Mixing in Photorefractive Crystal

2.1. Formation of Index Grating

2-WM in photorefractive crystal is briefly reviewed. In the case that the space-charge electric field is created by carrier diffusion, the phase of the induced index grating spatially shifts from the light interference fringe by 90 degrees [Kukhtarev, Markov, Odulov, Soskin and Vinetskii 1979]. The geometrical configuration of 2-WM using a photorefractive crystal is shown in Figure 1. Here, we assume the simplest scalar beam coupling in which the polar-izations of two beams are in a same direction. Due to the phase shift of the index grating by 90 degrees from the interference fringe, interference between the diffracted beam and the other transmitted beam becomes constructive, while, on the contrary, the other interference becomes destructive. Thus, amplification occurs for one beam which causes a loss of energy for another, resulting in energy transfer from the pump beam 2 to the signal beam 1.

2.2. Amplification Characteristics

From the coupled-mode theory, the expressions for the intensities of signal and the pump waves I_1 and I_2, respectively, are given by [Kukhtarev, Markov, Odulov, Soskin and Vinetskii 1979; Cronin-Golomb, Fischer, White and Yariv 1984; Fainman, Kiancnik and Lee 1987]

$$I_1(L) = rI_+(0)/\{r + exp(-\Gamma L)\} \tag{1}$$

$$I_2(L) = I_+(0)/\{1 + r \, exp(\Gamma L)\} \tag{2}$$

where $I_+(0)$ is the total power at $z = 0$, Γ is the gain coefficient, and r is the ratio of input signal beam intensity $I_1(0)$ to the pump beam intensity $I_2(0)$ at the input. Γ varies with the photorefractivity and externally applied electric dc field. In the limit without pump depletion that

$$1/r \gg exp(\Gamma L) \gg 1, \tag{3}$$

from Equation (1) for the signal beam is reduced to

$$I_1(L) = I_1(0) \, exp(\Gamma L). \tag{4}$$

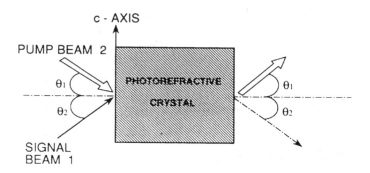

GEOMETRY OF TWO-WAVE MIXING

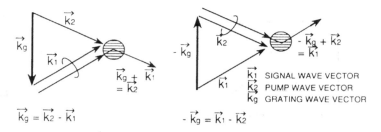

INTERACTION OF TWO BEAMS

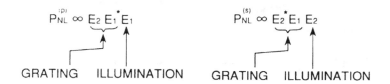

NONLINEAR POLARIZATION

Figure 1. Configuration of 2-WM using photorefractive crystal. Diagrams of index grating formation by the two wave-vectors and the nonlinear polarizations.

Photorefractive materials are now being revisited after the research was stagnated in the mid-1970s. Photorefractive crystals are categorized into three classes: ferroelectric crystals, paraelectric crystals, and semi-insulating semiconductors. In Figure 2, the figure of merit

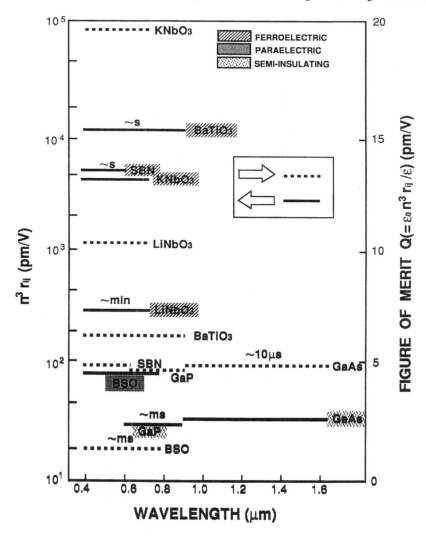

Figure 2. Pockel's coefficients by solid line and figure of merit Q by dashed line, of various classes of photorefractive crystals as a function of wavelength. The figure of merit is defined as described in the figure where ϵ_0, and ϵ are the permittivities in vacuum and the crystal. The line indicates a sensitive spectral range of the photorefractive effects.

Q of photorefractive crystals is plotted as a function of wavelength. The larger the figure of merit Q becomes, the deeper the index modulation gets. The response time, which is generally slow, conflicts with Q [Yeh 1987]. The menu of available crystals is still almost the same as the old one. Here, $BaTiO_3$ and $LiNbO_3$ are mainly used in our experiments.

3. Applications of Two-Wave Mixing

3.1. Logic Operations

In Figures 3(a) and (b), logic operations of NOT and NOR using photorefractive crystals are schematically shown. The truth tables of the logic operations are also shown. Here, PCW is featured. Waveguiding characteristics of PCW will be detailed later in Section 4. A lateral pump technique [Fischer and Segev 1989] is adopted. As shown in Figure 3, the clock beam serving as the pump source is laterally incident on the side of the waveguide. This offers a good tolerance for the launching alignment for the clock signal. In Figure 3(a), let the intensities of input signal beams 1(0) and 0(0) at $z = 0$ correspond to the logical values 1 and 0, respectively. In figure 4, the intensities of signal and clock (pump) beams are numerically plotted against the ratio r of signal beam intensity to clock beam intensity.

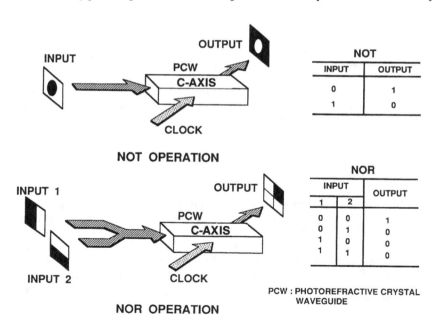

Figure 3. Schematics of NOT and NOR operations using PCW and the truth tables. Two possible directions of c-axis are indicated.

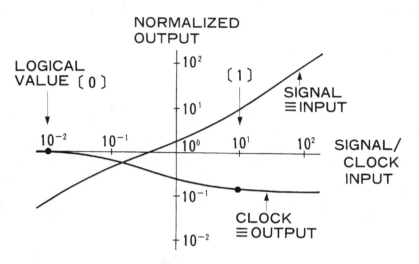

Figure 4. Intensities of signal pump beam in 2-WM against the ratio of input signal to input pump intensities. The ratios corresponding to logical values 0 and 1 are indicated. Here, $\Gamma L = 2.0$.

It is observed that if the intensities of input beams are appropriately set so that $1(0) \gg I_c(0) \gg 0(0)$ holds, being $I_c(0)$ the clock signal intensity at the launching point, the output clock signal can be approximated as

$$I_c \cong \begin{cases} 1(=I_c(0)) & \text{for } 0(0) \\ 0 & \text{for } 1(0). \end{cases} \tag{5}$$

In Figure 4, the dynamic range approximately 10 dB is retained for the case that $I_c(0)/1(0)$ and $0(0)/I_c(0)$ are around 10^{-2}. Thus, the logic operation NOT is executed. Note that a fraction of input power $0(0)(\neq 0)$ is necessary for the logical value 0 for the onset of 2-WM.

To execute NOR operation, a similar configuration as the one for NOT is used as shown in Figure 3(b). The intensities of two input signals for logical values 0 and 1 are set at $z = 0$ so that $1_1(0) + 1_2(0) \gg I_c(0) \gg 0_1(0) + 0_2(0)$ is satisfied where the subscript $i(i = 1, 2)$ denotes the input arm i. Then, the result for NOR operation is obtained as the output clock signal beam

$$I_c \cong \begin{cases} 1(=I_c(0)) & \text{for } 0_1(0) \text{ and } 0_2(0) \\ 0 & \text{for otherwise.} \end{cases} \tag{6}$$

3.2. Cross Connect

A novel scheme of optically controlled reconfigurable cross connect based upon a double color pumped oscillator (DCPO) [Fischer and Sternklar 1987] using a photorefractive crystal is shown in Figure 5. Assume that the beam with the wavelength of λ_0 bears the information, and the wavelength λ_j ($\neq \lambda_0$) of the other beams represents the routing address [Hashimoto, Fukui and Kitayama 1990]. Here, the 2×2 configuration is shown for simplicity. Suppose that the routing beam $R_i (i = 1, 2)$ is with the wavelength λ_j ($j = 2, 1$). R_i is incident on the crystal from the opposite side of the information-field beam IN_i with $\lambda = \lambda_0$. The data may take either 1-D or 2-D form. The pair of the incident beams IN_i and R_i serve as the pump sources for the onset of two 2-WMs, resulting in the generation of the output beam OUT_i with $\lambda = \lambda_0$ to the output port j and the other beam R_i' with $\lambda = \lambda_j$. This wave-mixing process originates from DCPO. As shown in the wavevector diagram of Figure 6, a common grating $\vec{k}_{gi}$ is formed both by the pump beam IN_i and self-generated beam and OUT_i and another pump beam R_i and self-generated R_i'. The wavevectors of self-generated beams $\vec{k}_{outi}$ and $\vec{k}_{ri}'$ are self-chosen through the oscillation such that a common grating $\vec{k}_{gi}$ ($=\vec{k}_{ini} - \vec{k}_{outi} = \vec{k}_{ri} - \vec{k}_{ri}'$) is created. A difference of $\Delta\lambda = \lambda_2 - \lambda_1$ between the routing beams makes the steering angle [Fischer and Sternklar 1987]

$$\Delta\theta/\Delta\lambda = -\sin \psi / \lambda_0. \tag{7}$$

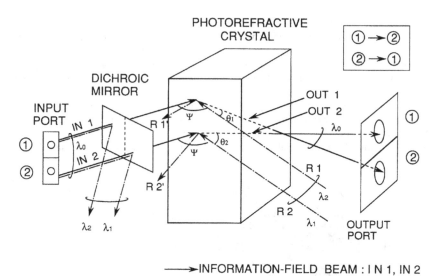

Figure 5. Configuration of optical cross connect based upon double color pumped oscillator (DCPO) using photoreflective crystal.

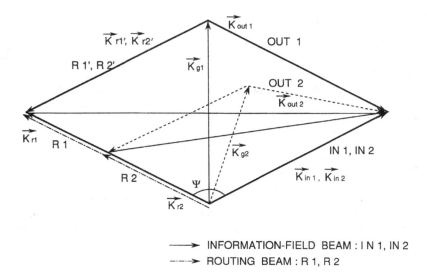

Figure 6. Wavevector diagram of double color pumped oscillator (DCPO).

For example, in the case that $\psi = 150$ degrees and $\Delta\lambda/\lambda_0 = 10\%$, $\Delta\theta$ becomes 2.9 degrees. This will provide a good isolation between the neighboring detectors at the output ports, and the separation of 500 μm is obtained at the distance of 1 cm away from the crystal.

It should be mentioned that besides DCPO, double color-birdwing phase conjugation (DC-BWPC) [Ewbank 1988] would be also applicable for the cross connect in the similar manner.

A rather long-term goal is an all-optical computer shown in Figure 7, consisting of optical processors and optical memories along with optical image amplifier. The optical cross connect plays an important role by routing data from processor to memory and vice versa. The use of photorefractive-crystal-based cross connect may allow routing 2-D data form in

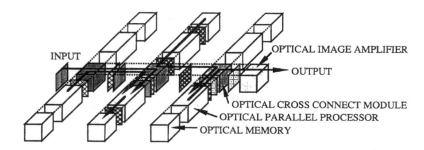

Figure 7. Schematic diagram of optical computer.

parallel. The crystal may also serve as the optical image amplifier. For the optical inter-connections in massively parallel computer architecture, reconfigurability of routing as well as high density of switching elements are at least requisite by taking into account 2-D image handling. The present cross connect technique has some attractive features such as rewritability on real-time basis and a high spatial resolution, equivalent of densely packed switching elements, although it is currently limited in terms of speed. The slow response may rule out ferroelectric crystals, such as $BaTiO_3$ and $LiNbO_3$ as the useful material in this application. However, photorefractive response times as short as 20 μsec have been observed in $GaAs$ [Klein 1984]. Furthermore, the advantage of semi-insulator semicon-ductors such as $GaAs$ and GaP [Kuroda, Okazaki, Shimura, Okamura, Chihara, Itoh and Ogura 1990] is that the high photorefractivity falls in an infrared spectral region as shown in Figure 2 where laser diodes are available for the light source.

4. Photorefractive Crystal Waveguides

4.1. Waveguide Geometry

PCW is a novel approach to enhancement of the nonlinear effects. In Figure 8, holographic recordings in bulk crystal and PCW are compared. In the bulk crystal, each ray of the reference plane wave interacts with the object beam only once at a specific location as shown in Figure 8(a). In the case of PCW in Figure 8(c), unlike the bulk crystal, the in-teraction of the waves is distributed throughout the waveguide. As the reference beam takes a zig-zag path, the beam can interact with the object beam each time it bounces back at the waveguide boundary. If the bulk crystal is laterally folded N times into the PCW geometry as shown in Figure 8(b), the volume-saving factor becomes N^2. As the reference beam hits the boundary M times with the period of ℓ as traversing the waveguide, this makes the interaction MN times over the entire distance L, resulting in the enhancement diffrac-tion efficiency by a factor of the length. Multiplicity due to the enhanced angular selec-tivity can also be increased in PCW. Moreover, a high optical energy density due to tight confinement of mode-field improves the energy efficiency.

Furthermore, a waveguide geometry also allows one to synthesize compact devices into any configuration such as a matrix array and bundle [Hesselink and Redfield 1988]. Thus, the fabrication difficulty of a large single crystal still existing for some special class of crystals can be circumvented. Although a loss of spatial information due to the limited aperture of waveguide could deteriorate image quality, however, the shortcoming can be overcome by compromising with the information storage capacity. For the alternative, phase conjugate mirror (PCM) is used to make it possible by compensating modal phase disper-sion which stems from the waveguide geometry to have all the guided modes bear image information and thus keep the storage capacity unsacrificed. The latter technique allows one to exploit to the full extent the large storage capacity of the waveguide. This will be elaborated in Section 4.2.

In the experiments below, $LiNbO_3$ crystal fibers are prepared [Yoshinaga, Kitayama and Oguri 1990]. The $LiNbO_3$ fibers prepared for this experiment are grown along the z axis by the resistance-heated pedestal growth downward method. Iron dopant concentration is expec-ted to be 0.005%. As the diameter is about 200 μm, the fiber used is highly multimoded. The

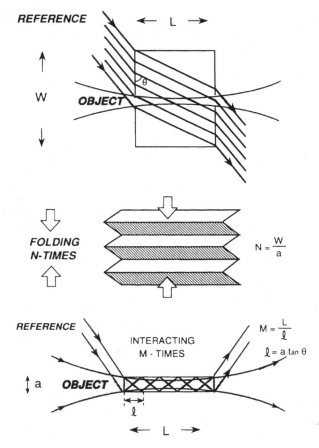

Figure 8. Interaction of object and reference beams in holographic recordings: (a) in bulk crystal and (c) in PCW. Multiple interaction of the two beams is illustrated by the folding geometry in (b).

Figure 9. Cross-sectional view of *LiNbO₃* fiber fabricated by the pedestal method.

length is 3 to 5 mm. Figure 9 shows a cross-sectional view of the fiber endface. As the cladding material is not deposited, the transmission loss of the fiber is relatively high, 1.4 dB. $BaTiO_3$ crystal waveguides are also prepared for the first time. It is cleaved out of a single bulk crystal. The PCW is 5 mm long, and the rectangular cross section is 1 mm².

4.2. Storage of Fourier Transformed Image

The $LiNbO_3$ crystal fiber shown in Figure 9 are used in this experiment. It is expected that the quality of the reconstructed image deteriorates since the aperture of the fiber is not large enough to record the whole Fourier spectrum of the image. In Figure 10, the reconstructed images of the single letter *J* for various focal lengths of the Fourier lens is shown. The incident angle of the reference beam to the fiber axis is 15 degrees. The shorter the focal length becomes, higher spatial frequency components can be recorded, resulting in the reconstruction with a higher fidelity. This is seen from Figure 10 that by comparing the case with $f = 1000$ mm in (a), the edge of the image with the case of shorter focal length $f = 300$ mm in (c) becomes clearer. However, the image in (d) for much shorter focal length shows a degradation of image fidelity. It is presumed that the high frequency components excite higher-order guided modes of the fiber. This causes the phase noise in the reconstructed image due to the modal phase dispersion. Therefore, once a fiber length and a diameter are given, a compromise has to be made to set the cutoff value for a spatial frequency to obtain the best fidelity. The theoretical analysis will be detailed in a separate paper.

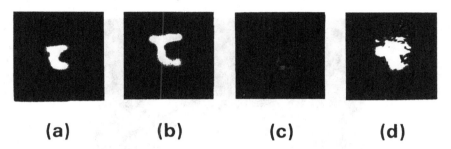

(a)   **(b)**   **(c)**   **(d)**

Figure 10. Experimental results of reconstruction of Fourier transformed images of letter J recorded in $LiNbO_3$ fiber. Focal lengths of Fourier lens are (a) $f = 1000$ mm, (b) 600 mm, (c) 300 mm, and (d) 50 mm.

A perfect recall of a hologram recorded in PCW is possible by using phase conjugation of the reference beam as shown in Figure 11. In the experiment, $BaTiO_3$ waveguides described

(a) Recording

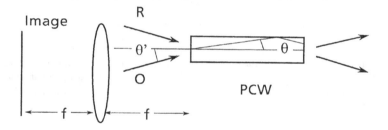

(b) Reconstruction

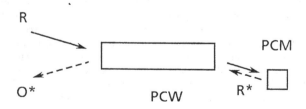

Figure 11. Schematic view of holographic storage of Fourier transformed image in PCW and reconstruction using PCM. (a) Recording and (b) reconstruction by generating phase conjugate of transmitted reference wave with PCM.

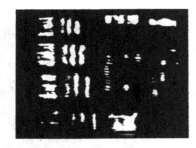 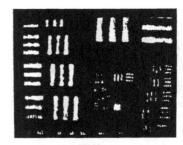

Figure 12. Experimental results of reconstruction. Image at the output end in the left and image at the input end restored through phase conjugation in the right. The incident angle of the image is 10 degrees.

in Section 4.1 are used. An argon laser ($\lambda = 514.5$ nm) is used. Here, a $BaTiO_3$ photorefractive crystal placed just behind the output endface of the PCW is used as the phase conjugate mirror PCM. As the distortion caused by the modal phase dispersion of PCW can be completely compensated by backpropagating phase conjugate reference wave on its original path, a perfect reconstruction should be expected at the input endface of the PCW. A similar method using phase conjugate mirror has been taken to restore the distortion of transmitted image [Yariv 1976]. In Figure 12, the experimental results for this recall using the Air Force Test Chart are shown. By comparing with the reconstructed images without using PCM observed at the output endface, the restoration of distortion is clear in the holograms reconstructed by PCM.

4.3. Optical Associative Memory

One will immediately find the application of PCW to optical implementation of neural networks [Rumelhart, McCleland and DDP Research Group 1986]. Photorefractive materials can serve as a storage device for synaptic weights in various networks due to the erasability and rewritability as a real-time holography [Owechko, Dunning, Marom and Soffer 1987; Abu-Mostafa and Psaltis 1987; Kitayama, Yoshinaga and Hara 1989]. In Figure 13, the experimental results of heteroassociative memory using PCW is shown. In Figure 14, the experimental setup for the associative memory using holographic recording in PCW is shown. Here, two PCWs are connected in tandem. The input endface of PCW is placed at the Fourier transform plane of a lens. The reference beam is incident obliquely on the waveguide, while, on the other hand, the object beam is incident on-axis. Fourier transforms of bright left-half and bright-right-half images are recorded in Fiber 1 with the reference beams $R1$ and $R2$, respectively, while in Fiber 2 letters U and J are recorded with $R1$ and $R2$, respectively. Imperfect or partial input image recalls the corresponding output letter.

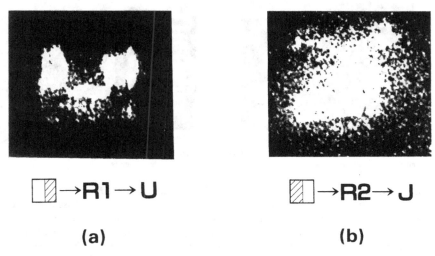

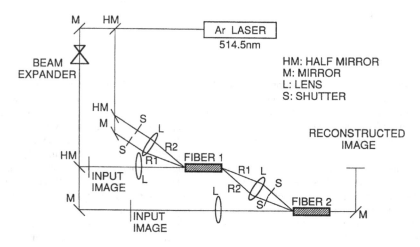

Figure 13. Experimental results of heteroassociative recall. The output from Fiber 2: (a) bright left-half image and (b) bright right-half image launched in Fiber 1 as an input image are recalled through the heteroassociations shown in the figure.

Figure 14. Experimental setup for holographic recording in PCW. The arrangement to perform associative memory using two PCWs aligned in tandem.

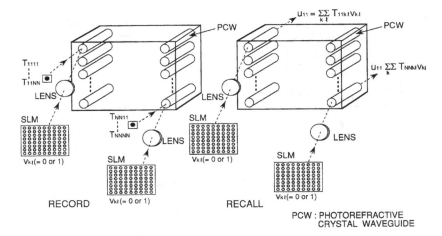

**SYNAPTIC WEIGHTS for IMAGE
USING 2-D PCW ARRAY**

Figure 15. Schematic diagrams for holographic recording and reconstruction in 2-D array of PCWs. Fourier transform of synaptic weights T_{ijkl} is recorded and the output for 2-D input image v_{kl} is recalled.

4.4. Two-Dimensional PCW Array and Its Application

The waveguide geometry allows one to synthesize compact devices into any configuration such as matrix array and bundle [Hesselink 1990]. Thus, the fabrication difficulty of a large single crystal, existing for some classes of crystals, can be circumvented. In Figure 15, the Fourier transformed image in 2-D array of PCWs is recorded. A strategy to optically store synaptic weights T_{ijkl} of neural network and obtain the system output u_{ij} for pixelized image v_{kl} is shown. The synaptic weight matrix $[T_{ijkl}]$ becomes 4th rank in the case of 2-D image [Hopfield 1982]. The size of the matrix to fully connect between $N \times N$ neurons in input layer and $M \times M$ neurons of output layer becomes extremely large $(MN)^2$, for example, 10^8 for 100×100 pixelized image. Therefore, there seems no realistic alternative to this approach to record such a huge matrix.

The isolation from the other surrounding PCW is crucial for densely packed 2-D array of PCWs because the recording and erasure processes of individual waveguides should not have any effect on the other ones. To accomplish the perfect isolation, optical energy confinement has to be achieved by depositing a cladding layer having lower refractive-index than that of the core outside the core. It should be mentioned that the test PCWs used did not have the cladding layer.

4.5. Waveguide Phase Conjugate Mirror

PCM is also observed in PCW. Generation of phase conjugate wave in optical fiber has been theoretically analyzed [Yariv, AuYeung, Fekete and Pepper 1978]. In Figure 16, the

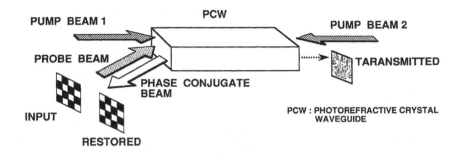

PCM BY FOUR-WAVE MIXING

Figure 16. Configuration of generating phase conjugate by four-wave mixing in PCW.

degenerate 4-WM using PCW, in which all the waves concerned have the same frequency, is schematically shown. As the two pump beams 1 and 2 counterpropagate each other, the phase matching condition of 4-WM is automatically achieved. Then, the probe beam launched in a waveguide will be replicated with high fidelity at the input of the probe beam. The phase conjugate wave is considered the diffractions caused by the two types of index grating, transmission and reflection gratings. The former is a replica of the interference fringe between the pump beam 1 and the probe beam, and the latter is the one between the pump beam 2 and the probe beam. The transmission gratings are considered to be dominant because they have a longer spatial frequency for which the photorefractive gain is large in ferroelectric crystals such as $BaTiO_3$.

The replacement of bulk-crystal-based PCM with PCW-PCM from the optical systems such as the one in Figure 11(b) will allow to simplify the system configuration. A preliminary experiment is conducted for the first time using a $BaTiO_3$ waveguide. From the experimental results shown in Figure 17, the generation of phase conjugate wave is confirmed due

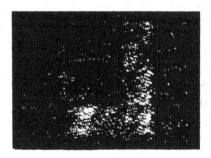 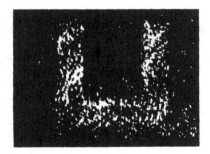

Figure 17. Experimentals results of phase conjugation by four-wave mixing in $BaTiO_3$ PCW. Distorted image at the ouptut end and restored image by phase conjugated wave are shown in the left and right, respectively.

to the fact that the fidelity of the letter U reconstructed by the PCM is higher than the transmitted image at the output endface. To improve the fidelity of the replicated pattern, there are some critical requirements for PCW-PCM. This will be a further study.

4.6. Two-Wave Mixing

2-WM can be achieved with the configuration using PCW shown in Figure 18 in which the c-axis is parallel to the waveguide axis. Assume that the signal beam (beam 1) excites lower order modes, and the pump beam (beam 2) excites higher order modes. In this configuration, the optical power couples from higher order mode to lower order mode since the gain coefficient of 2-WM is symmetric with respect to c-axis [Fischer and Sternklar 1987]. However, 2-WM occurs not only between the pump and signal beams but also between the guided modes excited by the signal beam. This degradates the fidelity of the image due to the modal dispersion of the guided modes bearing the image.

Let us consider another configuration of 2-WM using PCW shown in Figure 19(a). Since beams 1 and 2 exchange their roles in a manner that signal beam in one section acts like pump source in the next section and vice versa, the interaction of the two beams cannot transfer the energy unilaterally from one beam to another. To accomplish an efficient amplification by 2-WM in PCW, a practically promising method is to rotate periodically the c-axis by 180 degrees about the waveguide axis every period of the optical path of the ray as shown in Figure 19(b). Since this reversal changes the sign of the gain coefficient Γ in Equations (1) and (2), it is obvious that the interchange of the subscripts of the beams 1 and 2 in addition to Γ to $-\Gamma$ leads to the equations to their original forms. The signal beam is amplified efficiently by a particular pump guided mode whose period of the zig-zag path coincides with that of the c-axis reversal. As a result, the undesired coupling between the guided modes of the signal beam is suppressed, and thus the image amplification without distortion can be achieved. The experimental verification is now underway.

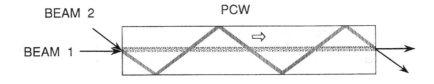

Figure 18. Configuration for two-wave mixing in PCW in which c-axis is parallel to the waveguide axis.

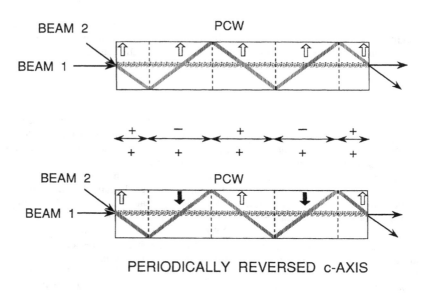

PERIODICALLY REVERSED c-AXIS

PCW : PHOTOREFRACTIVE CRYSTAL WAVEGUIDE
⇨ , ➡ : c-AXIS
    + : CONSTRUCTIVE INTERFERENCE FOR BEAM 1
    − : DESTRUCTIVE INTERFERENCE FOR BEAM 1

Figure 19. Configurations of two-wave mixing in PCW. (a) c-axis is perpendicular to the waveguide axis. (b) c-axis is periodically rotated by 180 degrees.

5. Conclusion

Novel applications of photorefractive effect to optical signal processing have been proposed and demonstrated. An emphasis has been put on the exploitation of massive parallelism of optical interconnection. First, the applications of two-wave mixing including logic operations and cross connect have been discussed. Next, photorefractive crystal waveguides (PCWs) and the applications have been focused on. Holographic storage of Fourier transformed image in PCW, two-dimensional array of PCW storage device, the phase conjugate mirror using PCW for the first time, and structural consideration of PCW for efficient two-wave mixing have been investigated both theoretically and experimentally.

Finally, it should be noted that as current technologies for optical signal processing are just too limited to mimic what a typical electronic counterpart can do. To realize the full potential of optical signal processing, advances will depend on the development of optical devices and materials.

Acknowledgments

The authors would like to thank Dr. S. Shimada and Dr. H. Ishio for their encouragement.

References

Y.S. Abu-Mostafa and D. Psaltis, *Scientific American* 256, p. 88, 1987.

H.H. Arsenault, T. Szoplik, and B. Macukow, *Optical Processing and Computing*, Academic Press: San Diego, 1989.

M. Cronin-Golomb, B. Fischer, J.O. White, and A. Yariv, *IEEE J. Quantum Electron.* QE-20, p. 12, 1984.

M.D. Ewbank, *Opt. Lett.* 13, p. 47, 1988.

Y. Fainman, E. Kiancnik, and S.H. Lee, *Opt. Eng.* 25, p. 228, 1986.

B. Fischer and M. Segev, *Appl. Phys. Lett.* 54, p. 684, 1989.

B. Fischer and S. Sternklar, *Appl. Phys. Lett.* 51, p. 74, 1987.

R.A. Fisher, *Optical Phase Conjugation*, Academic Press: New York, 1983.

J.W. Goodman, F.J. Leonberger, S.Y. Kung, and R.A. Athale, *Proc. IEEE* 72, p. 850, 1984.

P. Gunter and J.P. Huignard, *Photorefractive Materials and Applications*, Springer: Heidelberg, 1988.

M. Hashimoto, M. Fukui, and K. Kitayama, *IEEE Photonics Tech. Lett.* 2, p. 522, 1990.

R.W. Hellwarth, *IEEE J. Quantum Electron.* QE-15, p. 101, 1979.

L. Hesselink, *International J. of Optoelectronics* 5, p. 103, 1990.

L. Hesselink and S. Redfield, *Opt. Lett.* 13, p. 877, 1988.

J.J. Hopfield, *Proc. Matl. Acad. Sci.* USA 79, p. 2554, 1982.

K. Kitayama, H. Yoshinaga, and T. Hara, "Experiments of learning in optical perceptron-like and multilayer neural networks," *International Joint Conference on Neural Networks*, Washington, D.C., June 1989.

M.B. Klein, *Opt. Lett.* 9, p. 350, 1984.

N.V. Kukhtarev, V.B. Markov, S.G. Odulov, M.S. Soskin, and V.L. Vinetskii, *Ferroelect.* 22, p. 949, 1979.

K. Kuroda, Y. Okazaki, T. Shimura, H. Okamura, M. Chihara, M. Itoh, and I. Ogura, *Opt. Lett.* 15, p. 1197, 1990.

Y. Owechko, G.J. Dunning, E. Marom, and B.H. Soffer, *Appl. Opt.* 26, p. 1900, 1987.

D. Rumelhart, J.L. McCleland, and PDP Research Group, *Parallel Processing*, vol. 1, MIT Press: Boston, 1986.

P.W. Smith, *IEEE Circuits and Devices Magazine*, May 9, 1987.

A. Yariv, *Appl. Phys. Lett.* 28, p. 88, 1976.

A. Yariv, J. AuYeung, D. Fekete, and D.M. Pepper, *Appl. Phys. Let.* 32, p. 635, 1978.

P. Yeh, *Appl. Opt.* 26, p. 602, 1987.

H. Yoshinaga, K. Kitayama, and H. Oguri, *Appl. Phys. Lett.* 56, p. 1728, 1990.

Multidimensional Systems and Signal Processing, 2, 421–436 (1991)
© 1991 Kluwer Academic Publishers, Boston. Manufactured in The Netherlands.

Gaussian Wavelet Transform: Two Alternative Fast Implementations for Images

RAFAEL NAVARRO AND ANTONIO TABERNERO
Instituto de Optica "Daza de Valdés" (CSIC), Serrano 121. 28006 Madrid, Spain

Received February 26, 1991, Accepted April 5, 1991

Abstract. A series of schemes for pyramid multiresolution image coding has been proposed, all of them based on sets of orthogonal functions. Several of them are implementable in the spatial domain (such as wavelets), whereas others are more suitable for Fourier domain implementation (as for instance the cortex transform). Gabor functions have many important advantages, allowing easy and fast implementations in either domain, but are usually discarded by their lack of orthogonality which causes incomplete transforms. In this paper we quantify such effect, showing a Gaussian Wavelet Transform, GWT, with *quasiorthogonal* Gabor functions, which allows robust and efficient coding. Our particular GWT is based on a human visual model. Its incompleteness causes small amounts of reconstruction errors (due to small indentations in the MTF), which, however, are irrelevant under criteria based on visual perception.

Keywords. Gaussian wavelets, Gabor functions, image coding, completeness, space and frequency domains implementations

1. Introduction

Multiresolution signal decomposition is being commonly accepted as an optimal solution to processing, coding, and analyzing signals (see for instance [Rosenfeld 1984]). In this sense, the functional organization of the human visual system is based in this kind of image decomposition since individual cells in the cortex are tuned for specific bands of spatial frequencies and orientations [De Valois, Albrecht and Thorell 1982; Campbell and Kulikowski 1966]. In fact, the visual system performs a multiresolution, and also a *multiorientation*, parallel processing of images. Both neurophysiological [Marcelja 1980] and psychophysical [Daugman 1984] studies agree that the receptive fields, or frequency channels in the visual system are very accurately modelable by Gabor functions.

From a point of view of compact coding design, however, Gabor functions are not commonly accepted. Probably this is due to two reasons: first Gabor, in his early work [Gabor 1946] did not really address multiresolution pyramid coding. Second, and most important, Gaussian functions (wave packets) do not directly yield a complete (exact) mapping of the signal. Although this can be overcome by more or less sophisticated methods [Daugman 1988], many other complete transforms have been proposed, such as the Cortex Transform [Watson 1987]; the generalized Gabor scheme [Bastiaans 1981; Porat and Zeevi 1988]; wavelets [Morlet, Arens, Forgeau and Giard 1982; Mallat 1989]; QMF filters [Woods and Neil 1986; Simoncelli and Adelson 1990] etc. Most of those transforms (wavelets, generalized Gabor, QMF, etc.) have been designed to constitute a compact nonredundant coding, involving the computation of a new set of coefficients in a different orthogonal base. Other schemes are based on human vision models (the Cortex Transform, CT, or the Gaussian Wavelet Transform, GWT, proposed here) and basically consist of image decomposition into a set of frequency and orientation channels. Since some amount of redundancy seems to be an intrinsic feature of the image mapping in the visual system (which is multipurpose with a high performance), these schemes may involve some amount of redundancy. Thus, this last type of mapping yields a less compact, but more robust coding, and also shows other advantages; for instance, in order to recover the image it is not necessary to perform any inverse transform, but just to add the channels. As far as we know, it is not clear which among those schemes is the best, but this important number of different proposals indirectly means that image mapping is still an open problem, which is being crucial, e.g., to define standard codes for digital video and broadcast High Definition TV (see for instance [Watson 1990]).

The aim of this paper is to show that Gabor functions (Gaussian wavelets) are also good candidates for image coding, presenting interesting features such as robustness or the possibility of easy and fast implementations in either spatial or spatial-frequency domains. In what follows, we first propose, in Section 2, a Gaussian Wavelet Transform (GWT), which is based on a visual model and is designed to minimize incompleteness effects. The implementations in the Fourier and spatial domains are described in Sections 3 and 4, respectively. The effect of incompleteness are objectively quantified in terms of either the RMS (root mean square) reconstruction error in the spatial domain, or a Modulation Transfer Function, MTF in the Fourier domain, although subjective perceptual criteria can be more adequate in many applications [Gonzalez and Wintz 1977]. The GWT is also compared with other similar, but complete, transform (the Cortex Transform) in the discussion, Section 5.

2. Gaussian Wavelet Representation of Images

In this section we first describe briefly Gaussian wavelets, or Gabor functions, including the reciprocal representation in both spatial and frequency domains, namely receptive fields and channels respectively. Then, among all possible sets of Gabor functions, we shall make our choice based on visual models and image quality criteria.

2.1. Gaussian Wavelets

The Gaussian wavelet, or Gabor functions, is a complex exponential modulated by a Gaussian function. In two-dimensions (2-D), assuming radial symmetry and polar coordinates in the frequency domain, the analytical form in spatial variables is [Tabernero and Navarro 1991]:

$$g_{x_0, y_0, f_0, \theta_0}(x, y) = W \cdot g_{0, 0, f_0, \theta_0}(x, y) * \delta(x - x_0, y - y_0),$$

where

$$g_{0,0,f_0,\theta_0}(x,\ y) = \exp[-\pi a^2((x\cos\theta_0\ +\ y\sin\theta_0)^2\ +\ \Gamma^2(y\cos\theta_0\ -\ x\sin\theta_0)^2)].$$

$$\exp[i2\pi f_0(x\cos\theta_0\ +\ y\sin\theta_0)\ +\ \phi] \tag{1}$$

The shape of the Gabor function depends on the following parameters: W which is the gain or amplitude; a which is related to the radial bandwidth, and Γ which is the aspect ratio, while the labels x_0, y_0, f_0, and θ_0 account for the double localization in both space and frequency domains. The Gabor function is complex, having a symmetric (real) and antisymmetric (imaginary) parts. Following an analogy with the human visual system [Jones and Palmer 1987], we shall decompose each Gabor function into two *Receptive Fields* in the space domain, differing in their parity, $p = 0$ (even) or 1 (odd). This choice is not arbitrary because it implies very important computational advantages. Then each single receptive field is defined by five labels (x_0, y_0, f_0, θ_0 and p).

We have been dealing until now with Gabor functions in the spatial domain. However, one important advantageous feature of the Gaussian wavelets is their reciprocity in both domains (see Figure 1). This stems from the fact that the Fourier transform of a Gaussian is another Gaussian, which means that the functional form is basically the same in either the spatial or frequency domains. This will allow (next sections) the two alternative implementations in either domain to be equally easy. Such duality is broken in some way by our visual system, which in any case is ultimately the final receptor of images. It is well established that humans can make a very good job in localizing things in space, while their ability for spatial frequency analysis is relatively poor. Our visual system uses an *implementation* in the spatial domain performing a fine sampling of the space; however, this sampling is coarse in the frequency domain (this is imposed by the uncertainty relation) [De Valois 1982; Daugman 1984]. Implementations in the Fourier domain (where the spatial localization is in some way lost) would be more adequate in the hypothetical contrary case: a coarse

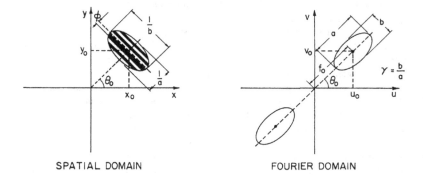

SPATIAL DOMAIN   FOURIER DOMAIN

Figure 1. Real component (even receptive field) of a Gaussian wavelet, or Gabor function, (left) and its Fourier transform (right), showing the main parameters of Equation 1.

sampling in space, but a fine Fourier analysis. In any case we shall talk about *Frequency Channels* when dealing with the Fourier domain, and about *Receptive Fields* as regards to the spatial domain.

2.2. The Set of "Logons"

Following Gabor [Gabor 1946], in order to represent a signal one can decompose it into a set of elementary quanta of information or logons. Image representations in space (pixels) or by Fourier analysis (spatial frequencies) are just special cases. There is a lot of evidence showing that such representations of images are not the best. Depending on the application one could choose any intermediate case allowing a more or less coarse (or fine) local spectral analysis [Jacobson and Wechsler 1988]. The use of Gabor functions will insure an optimum packing of information (minimizing the uncertainty relation) and the reciprocity in both domains. Unfortunately, Gaussian wavelets present a drawback: due to their shape they do not constitute a complete set. However, by quantifying the reconstruction error, it is possible to find a *quasicomplete* set of Gabor functions with a minimum error, which as we will show is perceptually irrelevant.

Like most of the proposed transforms, our scheme is based on a linear scale in the spatial domain; i.e., a uniform distribution of receptive fields, and a logarithmic sampling in the Fourier domain; i.e., frequency channels distributed in octaves ($\log_2$). This permits multiscale pyramid implementations [Burt and Adelson 1983] with self-similar wavelets: all the wavelets can be obtained by translating, rotating or scaling one of them. The octave distribution of channels has shown to be optimal since the spectra of natural images tend to show exponential type decays [Navarro, Santamaría and Gómez 1987; Field 1987]. This means that the energy per frequency channel tends to be constant.

The additional criteria to design the set of *logons* (the shape and number of wavelets) have been in our case [Tabernero and Navarro 1991]: known features of the human visual system, optimum coverage of both domains and simplicity. The resulting set is composed by four orientation channels ($\theta_0 = 0^0, 45^0, 90^0$ and 135^0) and four frequency channels—and a low-pass residual [Burt and Adelson 1983]. The peak frequencies (f_0) of the four channels are: half the Nyquist frequency ($f_N/2$) for the highest frequency channel, and then halving for the next lower octave channel ($f_N/4$, $f_N/8$ and $f_N/16$). The radial bandwidth has been set to one octave, $a = 0.71 \cdot f_0$ in Equation 1, which yields a nearly optimum coverage of the radial frequency axis. The aspect ratio $\Gamma = 1$ allows circular symmetry, while the resulting angular bandwidth (0.71 radians) is not far from the optimum angular covering of the Fourier domain (for 4 channels is $\pi/4 = 0.785$). Finally, the localization of the receptive fields in space will be given by sampling requirements; sampling in a square grid, with an interval given by the Nyquist frequency of the corresponding channel. Figure 2a shows the basic 4×4 Gabor functions in the spatial domain, and Figure 2b the recovering of the Fourier domain by this set. This last figure clearly illustrates the nature of the incompleteness of the GWT, since the recovery of the Fourier domain is not complete. This causes a nonflat MTF: apart from the loss of high frequencies mainly in the corners (high frequency residue; [Burt and Adelson 1983; Watson 1987]), what is characteristic of the Gabor functions are the indentations between adjacent channels (this is further illustrated in

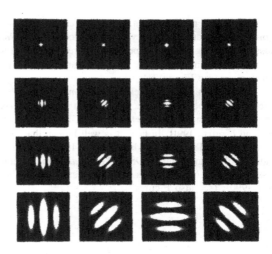

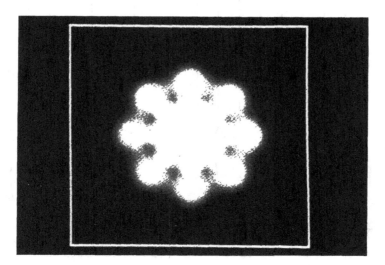

Figure 2. The 4 × 4 basic Gabor functions of the GWT: (a) The (even) Receptive Fields in the space domain; (b) Coverage of the Fourier domain by the Frequency Channels.

Figure 4). This makes the GWT be incomplete since a complete scheme would provide a uniform recovery of both domains.

Once the set of logons is defined, the image representation is obtained by the expansion into these elementary signals. Any implementation will involve the computation of the inner product of the signal with every Gabor function. The resulting numbers are samples of the image in a joint spatial/spatial-frequency domain [Jacobson and Wechsler 1988]. The logarithmic Gaussian mapping has the important feature that the joint size of logons is constant and optimal (minimum uncertainty relation means optimum packing of information) in the joint domain (see Figure 3).

An important aspect of this kind of scheme is that one only needs the even real part of the Gabor functions to encode the image. However, in image analysis, the performance can be highly improved by using the complete transform, including even and odd receptive fields [Tabernero and Navarro 1991]. In fact, it is well known that the simple neurons in the visual cortex are arranged in couples, differing just because their receptive fields are in phase quadrature [Pollen and Ronner 1983]. This *redundancy* seems to be inherent to the visual process: the even receptive fields are bar detectors, while the odd ones are edge detectors [Watson 1990]; (see also Figure 7).

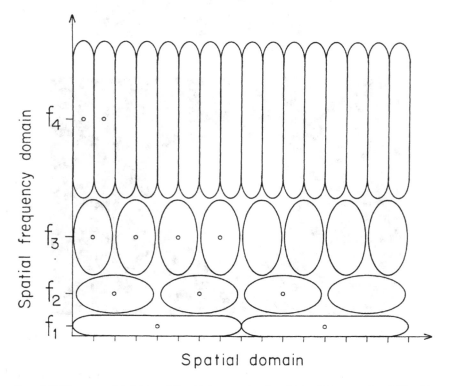

Figure 3. GWT coverage of the joint spatial/spatial frequency domain, with the minimum number of Gabor functions (logons), all of them having the same area (constant and maximum uncertainty relation).

3. Implementation in the Fourier Domain

As far as we know, the implementation in the Fourier domain has been common in this kind of transform [Watson 1987], since on the one hand the frequency channels are straight-forwardly implementable as linear Fourier filters (in our case Gaussian windows in the frequency domain, Figure 2b), and on the other hand, by using a Fast Fourier Transform algorithm, and a pyramid implementation [Watson 1987], the computer time is quite reasonable.

Most of the details of this implementation are the same as in similar schemes [Watson 1987]. The unique but important question is that the Gaussian channels do not allow a complete and uniform covering of the Fourier domain. This can be appreciated in Figure 4. This figure shows the modulation transfer function, MTF of the Gaussian Wavelet

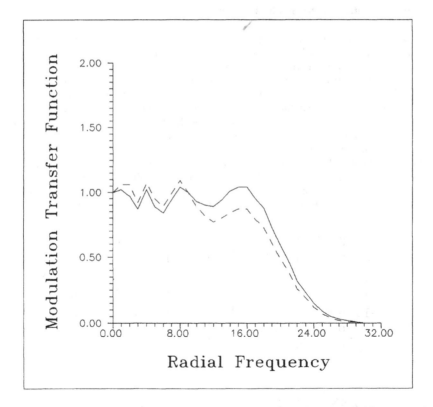

Figure 4. Modulation Transfer Function of the Fourier domain GWT implementation. The dashed line corresponds to the case of constant gain for the different channels, and the continuous line is after equalization by assigning different gains to the frequency channels (gains are 0.93, 0.93, 0.93 and 1.15 respectively).

Transform GWT, with, and without, assigning different weights (gains) to the different channels. The reason to assign different weights to the channels is to equalize (flatten) the MTF as much as possible; i.e., to minimize the reconstruction errors. The method used to find the gains, in both spatial and Fourier implementations, was empirical. The MTF was obtained using a random white noise as test image, as the ratio of the Fourier spectra: that of the recovered image after the GWT, versus that of the original. As can be appreciated, it is not possible with Gausian functions to obtain a completely flat MTF. This explains why the GWT is not complete, in the sense that we do not exactly recover the original. However, we will show that while objective criteria (RMS error in space, or MTF in frequency), give noticeable differences with respect to the original, under subjective visual criteria it is difficult to distinguish between the original and the reconstruction.

4. Implementation in the Spatial Domain

The self-similarity of the Gabor functions, along with their Gaussian shape, allows easy pyramid implementations in either spatial or frequency domains. Although any filtering operation can be equally performed in both domains by the convolution theorem, the Fast Fourier Transform (FFT) algorithms usually are more appropriate to implement the filtering in the Fourier domain. However, adequate filter design, with small filter convolution masks, can match, or even improve, the computing time when compared with a FFT implementation. For instance, 3×3 filter masks are widely used in real time image processing, including low-pass Gaussian filters. For this purpose we started designing a couple (even and odd) of filter masks corresponding to the highest frequency, horizontal $\theta = 0^0$ channel. This consisted of finding the minimum number of samples needed to reasonably represent the shape of that Gabor function. As a result of this tradeoff, the receptive fields were implemented in 7×7 convolution masks [Tabernero and Navarro 1991]. The resulting filter masks, shown in Figure 5, permit a fast implementation. The computing time for

$$\begin{pmatrix} 0 & -3 & 0 & 13 & 0 & -3 & 0 \\ 0 & -8 & 0 & 16 & 0 & -8 & 0 \\ 0 & -13 & 0 & 22 & 0 & -13 & 0 \\ 0 & -16 & 0 & 26 & 0 & -16 & 0 \\ 0 & -13 & 0 & 22 & 0 & -13 & 0 \\ 0 & -8 & 0 & 16 & 0 & -8 & 0 \\ 0 & -3 & 0 & 13 & 0 & -3 & 0 \end{pmatrix} \quad \begin{pmatrix} 3 & 0 & -8 & 0 & 8 & 0 & -3 \\ 6 & 0 & -11 & 0 & 11 & 0 & -6 \\ 8 & 0 & -15 & 0 & 15 & 0 & -8 \\ 10 & 0 & -16 & 0 & 16 & 0 & -10 \\ 8 & 0 & -15 & 0 & 15 & 0 & -8 \\ 6 & 0 & -11 & 0 & 11 & 0 & -6 \\ 3 & 0 & -8 & 0 & 8 & 0 & -3 \end{pmatrix}$$

EVEN FILTER   ODD FILTER

Figure 5. Filter masks of 7×7 pixels used in the spatial domain implementation. They correspond to the maximum frequency $f_N/2$, vertical $\theta = 0^0$, even and odd receptive fields.

the spatial implementation is roughly proportional to $7^2 \cdot N^2$ (N being the number of rows in a square image), versus $k \cdot N^2 \cdot \log_2 N^2$ (typical values for k are between 2 and 3) of the fast Fourier implementation. In consequence, for typical image sizes both implementations are approximately equivalent in computing time.

Once the basic couple of receptive filters is designed, the complete set is obtained by rotating and/or scaling operations. While the four orientations are in fact obtained by rotating the originals, a pyramid implementation is more adequate for scaling: instead of magnifying the filters, which would imply much more computing time, the pyramid implementation applies the same set of filters to successive compressed versions of the image. Each compressed version is obtained by first applying a low-pass Gaussian filter (5×5 mask) to the image in order to avoid aliasing, and then a decimation by a factor of 2 in both dimensions [Burt and Adelson 1983]. This scheme is typical in multiresolution methods [Rosenfeld 1984]. Figure 6 shows the result of the GWT expansion. The image is recovered by adding only the even logons (a), while the odd part (b) is equivalent to a set of *edge detectors*.

As in the Fourier domain implementation, the effect of incompleteness is that the GWT is not exact. Furthermore, undersampling of the receptive fields causes a broadening of the Fourier channel, which affects the MTF. This can be appreciated in Figure 7. This figure shows the MTF of the spatial domain GWT (computed from a white noise image), with and without assigning different weights, or gains, to the different channels. The equalized version is obtained with gains 0.45, 0.60, 0.65, and 0.90 for the four frequency channels, f_1 to f_4 respectively (the gain of the low frequency residual, 1, is unchanged). The assignment of different weights to the channels allows the equalization of the two implementations.

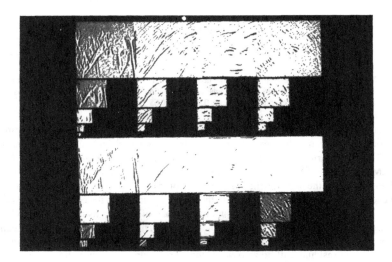

Figure 6. Result of the complete GWT. The grey scale is proportional to the value of each logon. Note that the even receptive fields (up), behave as bar detectors, and are enough to code and reconstruct the image. The odd receptive fields are edge detectors (bottom).

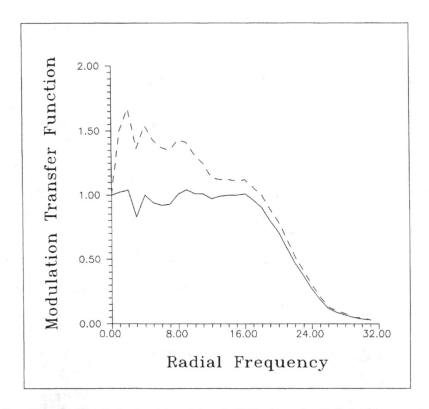

Figure 7. Modulation Transfer Function of the spatial domain GWT implementation. The dashed line corresponds to the case of constant gain for the different channels, and the continuous line is after equalization by assigning different gains to the frequency channels (gains are 0.45, 0.60, 0.65 and 0.90 respectively).

The matching is not total anyway, since the two implementations involve different computing errors, sampling, quantization, etc. Figure 8 shows (a) the original image, and reconstructions: (b) from Fourier domain and (c) from spatial domain implementations of the GWT. The RMS normalized errors of the reconstructions are 13% in both cases. Most of this RMS error (about the 11% for the CT. See Section 5) comes from disregarding the high frequency residual band, corresponding to the energy contained in the corners of the Fourier domain [Watson 1987]. In fact, the visual appearance of the reconstructions resembles a very slightly low-pass filtered version of the original. The high frequency residual can be incorporated to the GWT, but its value is usually going to be negligible. Moreover, this high frequency residual will be zero when the acquisition system uses both conventional optical systems and adequate sampling.

Figure 8. Original image (a) and reconstructions from Fourier domain (b), and spatial domain (c) implementations of the GWT.

5. Discussion

In this section we shall compare our GWT with a similar but complete transform. For this purpose, among the different published schemes, we chose the Cortex Transform, CT [Watson 1987], because this is probably the one presenting more similar features with respect to our GWT (4 × 4 channels, also based on visual models, etc.). The main difference is that the CT is complete; also the CT is difficult to implement in the spatial domain. In what follows, we use a version of the CT with exactly the same frequency, and orientation channels as our GWT.

5.1. Comparison Under Objective and Subjective Visual Criteria

Figure 9 shows two reconstructions from the CT (a) and the GWT (b) respectively. In both cases the high frequency residual has been neglected, which causes most of the error. The RMS error is slightly smaller in the CT case: 11% in the CT (which is only due to neglecting the high frequency residual), versus 13% in the GTW; while the visual appearance is almost identical. The effect of incompleteness is objectively quantizable in the Fourier

Figure 9. Comparison of the reconstructions obtained with the Cortex Transform, CT (a), and the Gaussian Wavelets Transform GWT (b). In both cases, the peak frequencies and orientations are the same, and the high frequency residual has been neglected.

domain as little indentations in the MTF of the transform or as a higher RMS error in the reconstruction. However, since the final receptor of images is usually going to be the human eye, subjective perceptual criteria are more adequate. In Figure 9, the GWT reconstruction could be better classified by several observers, although most of them would find it very difficult to establish a difference. This result points out that although Gabor functions have been previously discarded because of their lack of completeness, the GWT proposed here is *quasicomplete*; i.e., the effect of incompleteness is small in terms of RMS error, and irrelevant under perceptual criteria.

The redundancy inherent in using an incomplete representation has been also argued against Gabor functions. For image compression applications, orthogonal transforms should give a better result than those nonorthogonal; i.e., a more compact coding with smaller entropy or bit/pixel rate, since minimum entropy also means minimum redundancy [Gonzalez and Wintz 1977]. To assess that redundancy, we have computed and compared the entropy of the different frequency channels of Figure 6, with that equivalently obtained by the CT. The resulting values for the higher frequency channel are 4.4 and 4.3 respectively, which involves only a 2% of increment, while for the other channels there is not a noticeable difference (values are 4.6, 5 and 5.2 respectively. Note that these values of entropy are relatively high because they have been computed using the original quantization in 256 linear levels. By an adequate quantization and coding, the entropy can be highly reduced). In order to complete this comparison we also need to know the 2-D (area) bandwidth ratio between the GWT and CT frequency channels, because the total amount of bits required to code a channel is proportional to the entropy of that channel times its bandwidth. This has been estimated in the following way: first the minimum box in the Fourier domain containing a 99% of the energy of a CT channel has been determined, and second, the percentage of energy of the equivalent GWT channel inside that box has been computed. The result is 90% approximately, which once again does not imply a big difference. Actually this means a small increment in the amount of redundancy in the GWT with respect to the CT, which, as we shall show in Section 5.2, increases the robustness of the coding. Additionally, the reciprocity of Gaussian wavelets under Fourier transformation, makes them specially appropriate as a base of functions in joint representations. Moreover, the smoothness of Gaussian windows is very well suited to avoid edge effects, aliasing and ringing artifacts, etc., which are typical in Fourier filtering. As we shall show, this property, along with some amount of information redundancy, is important in robust coding.

5.2. *Comparison in Terms of Robustness of Coding*

One of the main applications of these kinds of transforms (Cortex, QMF, wavelets, etc.) is image coding. There is strong evidence suggesting that the human visual system also uses a similar scheme [Marcelja 1989; De Valois et al. 1982; Daugman 1984] and it has been shown that similar transforms allow high rates of data compression [Daugman 1988; Watson 1987]. Also the visual system makes an important use of redundancy: for instance the arrangement of neurons in couples with phase quadrature involves redundancy, but it is very important in shape and edge detection, or texture recognition. Many tasks can be more efficiently performed with this visual mapping of images, but also this mapping is well suited for robust coding.

Figure 10. Image reconstructions with partial losses of frequency channels. The first row (up) corresponds to the CT and the second (down) to the GWT. The first column (left) shows the reconstruction from only three frequency channels (missing the high frequency channel), and the second and third columns to reconstructions from two and only one channels, respectively.

Probably the most important features of good coding are robustness and efficiency. In particular, the robustness under partial information losses (or errors), which could be random or even systematic, is critical [Watson 1990]. Pyramid multiresolution coding is robust under high frequency losses. This has been tested, and the result is shown in Figure 10. The first row presents image reconstructions from the CT, with progressive loss of high frequency channels: with only three channels (first column left, 19%), and with two (second column $E_{RMS} = 26\%$), and just the lowest frequency channel (third, $E_{RMS} = 34\%$). The second row corresponds to the GWT in the same conditions ($E_{RMS} = 20\%$, 26% and 35%, respectively). In spite of these relatively high values of the RMS error, this figure illustrates the robustness of this kind of representation. Our eye can perform a very good job in face recognition, even when just the two lower frequency channels are available (a small subset, 6% approximately, of the total number of logons). This also could exlain to some extent why this coding is convenient in human vision, since people with low visual acuity (myopic, amblyopics, etc.) can recognize objects, and perform most visual tasks.

Figure 10 also illustrates a very important point. In the case of any kind of information losses, or errors, no coding is complete. In consequence, robustness is not at all dependent on completeness. If we compare the first and second rows of Figure 10, we can observe some artifacts, *ringing*, which are more clear in the first row, CT. Once again, although the RMS error is slightly smaller for the CT (11%, 19%, 26% and 34% in the CT versus 13%, 20%, 26% and 35% in the GWT), visual criteria will not make the same difference. Besides, in the case of partial losses (compare the first and second couples of pictures) the GWT presents slightly smoother artifacts, which means that under visual criteria, the GWT would be preferable. This fact is further illustrated in Figure 11, representing orientation channels losses.

Figure 11. Image reconstructions with partial losses of orientation channels. The first row corresponds to the CT and the second to the GWT. The first column shows reconstructions without a single orientation θ_{45} in only one frequency f_4 (high frequency). The second and third columns show reconstructions with one orientation channel missing for all frequencies (θ_0 and θ_{135} respectively).

Figure 11 shows reconstructions with orientation channel losses: the first column (left) corresponds to loosing a single orientation channel ($\theta_0 = 45^0$) only for the frequency band f_3; the second and third columns correspond to orientation losses (θ_0 and θ_{135} respectively) for all frequencies. The first row (top) corresponds to the CT and the second (bottom) to the GWT. Apart from some more or less curious effects (noise removal, etc.), this is a significant example in which the CT is subject (on average) to more visual artifacts, the GWT appearing to be more robust. This is because when an orientation channel is missing in the CT, which is less redundant, a deep and sharp hole appears in the Fourier domain. On the contrary, with Gabor functions, that hole is much less deep since it is partially filled by the adjacent channels, and of course it has much smoother edges. This is a very important advantage, because in these examples, the partial filling of holes in the GWT will allow image restoration (by inverse or Wiener filtering methods, etc.), while the deep holes in the CT are very difficult to restore.

Another important aspect of the GWT is that it is well suited for parallel optical implementations. In this sense, an optical wavelet transform has been recently proposed by [Freysz, Pouligny, Argoul and Arneodo 1990]. With an adequate design of optical filters, it would be amenable to completely parallel optical implementation; i.e., by multiplying the object by a high frequency square array of dots, many replicas of the object spectrum can be simultaneously obtained allowing to apply several filters at the same time.

6. Conclusion

Gaussian wavelets (Gabor functions) present a series of important advantages with respect to other wavelets, at the cost of incompleteness. Many authors have argued against Gabor functions for their lack of orthogonality and completeness, and proposed other alternative complete schemes and orthogonal bases of functions. However, we have shown that by a careful choice of the Gabor functions, in order to optimize the recovery of the frequency domain, the effect of incompleteness of the GWT is small, and perceptually almost insignificant. Moreover, the GWT proposed here can be considered in practice quasicomplete. The advantages are very important (most of them were already shown by Gabor in 1946): minimizing the uncertainty relation (optimum packing of information); formal reciprocity, which makes implementations possible and easy in either domains, and robustness under information losses, minimizing artifacts, etc. A final advantage is that the GWT is also based on human vision. Since the human visual system shows a high performance in most aspects, which has not been improved, or even reached, by artificial systems, image processing environments based on visual models are expected to be well suited for artificial vision.

Acknowledgments

This research has been supported by the Comisión Interministerial de Ciencia y Technología, Spain, under grant P18-88/88-0198 (PRONTIC).

References

M.J. Bastiaans, "A sampling theorem for the complex spectrogram, and Gabor's expansion of a signal in Gaussian elementary signals," *Opt. Engineer.* 20, pp. 594–598, 1981.

P.L. Burt and E.H. Adelson, "The laplacian pyramid as a compact image code," *IEEE Transactions on Communications* COM-31, pp. 532–540, 1983.

F.W. Campbell and J.J. Kulikowski, "Orientation selectivity of the human visual system" *J. of Physiology* 187, pp. 437–445, 1966.

J.G. Daugman, "Spatial visual channels in the Fourier plane," *Vision Research* 24, pp. 891–910, 1984.

J.G. Daugman, "Complete discrete 2-D Gabor transform by neural networks for image analysis and compression," *IEEE Transactions on Acoustic Speech and Signal Processing* ASSP-36, pp. 1169–1169, 1988.

R.L. De Valois, D.G. Albrecht, and L.G. Thorell, "Spatial frequency selectivity of cells in macaque visual cortex," *Vision Research* 22, pp. 545–559, 1982.

D.J. Field, "Relation between the statistics of natural images and the response properties of cortical cells," *J. of the Optical Society of America A* 4, pp. 2379–2394, 1987.

E. Freysz, B. Pouligny, F. Argoul, and A. Arneodo, "Optical wavelet transform of fractal aggregates," *Phys. Rev. Lett.* 64, pp. 745–748, 1990.

D. Gabor, "Theory of communication," *J. Inst. Elect. Eng.* 93, pp. 429–457, 1946.

R.C. Gonzalez and P. Wintz, *Digital Image Processing.* Addison-Wesley: London, 1977.

L.D. Jacobson and H. Wechsler, "Joint spatial/spatial-frequency representations," *Signal Processing* 14, pp. 37–68, 1988.

J. Jones and L. Palmer, "An evaluation of the two-dimensional Gabor filters model of simple receptive fields in cat striate cortex," *J. of Neurophysiology* 58, pp. 538–539, 1987.

S.G. Mallat, "A theory of multiresolution signal decomposition: the wavelet representation," *IEEE Transactions on Pattern Analysis and Machine Intelligence* PAMI-11, pp. 674–693, 1989.

S. Marcelja, "Mathematical description of the response of simple cortical cells," *J. of the Optical Society of America* 70, pp. 1297-1300, 1980.

J. Morlet, G. Arens, I. Forgeau, and D. Giard, "Wave propagation and sampling theory," *Geophysics* 47, pp. 203-236, 1982.

R. Navarro, J. Santamaría, and R. Gómez, "Automatic log spectrum restoration of atmospheric seeing," *Astronomy and Astrophysics* 174, pp. 334-351, 1987.

D.A. Pollen and S.F. Ronnen, "Visual cortical neurons as localized spatial filters," *IEEE Transactions on Systems, Man and Cybernetics* SMC-13, pp. 284-302, 1983.

M. Porat and Y.Y. Zeevi, "The generalized Gabor scheme for image representation in biological and machine vision," *IEEE Transactions on Pattern Analysis and Machine Intelligence* PAMI-10, pp. 452-468, 1988.

A. Rosenfeld, *Multiresolution Image Processing and Analysis*. Springer-Verlag: New York/Berlin, 1984.

E.P. Simoncelli and E.H. Adelson, "Nonseparable extensions of quadrature mirror filters to multiple dimensions," *Proc. of the IEEE* 78, pp. 652-664, 1990.

A. Tabernero and R. Navarro, "Performance of Gabor funcitons for texture analysis," *IEEE Transactions on Pattern Analysis and Machine Intelligence* PAMI (submitted).

A.B. Watson, "The cortex transform: rapid computation of simulated neural images," *Computer Vision, Graphics and Image Processing* 39, pp. 311-327, 1987.

A.B. Watson, "Perceptual-components architecture for digital video," *J. of the Optical Society of America A* 7, pp. 1943-1954, 1990.

J.W. Woods and S.D. O'Neal, "Subband coding of images," *IEEE Trans. on Acoustic Speach and Signal Procesing* ASSP-34, pp. 1278-1288, 1986.

Multidimensional Systems and Signal Processing, 2, 437–440 (1990)
© 1991 Kluwer Academic Publishers, Boston. Manufactured in The Netherlands.

Contributing Authors

M.A. Fiddy attended the University of London. He obtained a first class degree in Physics in 1973, and a Ph.D. in 1977. His thesis topic was the application of the theory of entire functions in optical scattering. He worked as a post-doctoral research assistant in the Physics Department at Queen Elizabeth College, University of London, and in the Department of Electronic and Electrical Engineering at University College London. These positions concerned the development of optical techniques for image recovery and optical fiber sensors, respectively. In 1979 he was appointed Lecturer in Physics at Queen Elizabeth College, moving to Kings College London in 1983. During part of 1982, he was visiting associate professor at the Institute of Optics, University of Rochester, and in 1985–86 held a similar position in the Mathematics Department at the Catholic University of America, Washington, D.C. In September 1987 he moved to the University of Lowell where he is a professor of electrical engineering.

Dr. Fiddy was on the Editorial Board of the journal *Inverse Problems* (Institute of Physics) from 1984 to 1990. He is currently on the editorial boards of *Waves in Random Media* (Institute of Physics), *Multidimensional Systems and Signal Processing* (Kluwer) and *Optical Computing and Processing* (Taylor and Francis). He has worked for many years in the field of scattering and inverse problems, spectral estimation and optical signal processing; he has over 60 refereed publications and over 90 conference publications.

H.J. Caulfield is the founder and director of the Center for Applied Optics at the University of Alabama in Huntsville. His work in areas such as optical computing, holographic, and metrology have spanned almost thirty years and has produced four books, numerous book chapters, and hundreds of technical papers. He has been honored by a number of technical societies and is both the author of and the subject of numerous popular articles as well. He is currently on the editorial boards of a dozen technical journals and has been editor of one  and was at one time editorial board member of two others. He and his wife Jane operate Far Out Farm in Tennessee which has produced a number of champion Romney sheep including the one joining Dr. Caulfield in the photograph.

David Mendlovic received the B.Sc. degree in Electrical Engineering from Tel-Aviv University, Israel in 1987.

He joined the Optical Image Processing Laboratory in the Department of Electrical Engineering—Physical Electronics, Tel-Aviv University, Israel in 1987 as a research student. He just completed his Ph.D. thesis on invariant pattern recognition.

Naim Konforti received the B.Sc. degree in Electrical Engineering from Technion Israel in 1960, and M.Sc.E.E. degre from the same Institute in 1975.

Since 1975 he has been working in the Optical Image Processing Laboratory in the Department of Electrical Engineering—Physical Electronics, Tel-Aviv University, Israel as a researcher and laboratory engineer.

Emanuel Marom received the B.S. and M.S. degrees in Electrical Engineering from Technion, Haifa, Israel, in 1957 and 1961, respectively, and the Ph.D. degree in Electrical Engineering from the Polytechnic Institute of Brooklyn, NY, in 1975.

Since 1972, he has been with the Tel Aviv University Faculty of Engineering, where he is a Professor of Electro-Optics at present. His research interests include various aspects of holography and diffraction, optical correlation, image processing, integrated optics, fiber optics components for signal processing applications and neural networks.

Dr. Marom is a Fellow in the Optical Society of America and a Senior Member of IEEE.

Dejan Pantelic received B.Sc., M.Sc. and Ph.D. degrees from the University of Belgrade in 1980, 1983 and 1990 respectively. Presently he is employed in the Institute of Physics in Belgrade. He is interested in optical signal processing, optical computing and optical memories.

Ken-ichi Kitayama received the B.E., M.E., and Dr. Eng. degrees in communication engineering from Osaka University, Osaka, Japan, in 1974, 1976, and 1981, respectively. In 1976, he joined the NTT Laboratories, where he was engaged in research work on optical fibers including transmission characterization and nonlinear fiber-optics. In the academic year of 1982–1983, he studied semiconductor lasers and integrated optics at the University of California, Berkeley as a visiting researcher. In 1987, he initiated the research on optical signal processings at the NTT Transmissions Systems Laboratories. He published approximately eighty papers in the established journals in English.

Dr. Kitayama received the 1980 Young Engineer Award from the Institute of Electronics and Communication Engineers of Japan, and 1985 Paper Award of Optics from the Japan Society of Applied Physics.

Fumihiko Ito received the B.S. degree in electrical engineering from University of Tokyo, Tokyo, Japan, in 1985. He joined the Electrical Communications Laboratories, Nippon Telegraph and Telephone Corporation, Ibaraki, Japan, in 1985. Since 1985, he has been engaged in research work on optical devices in subscriber optical fiber loops. His current interest includes photorefractive crystals and their applications to optical parallel signal processings. In 1990, he studied at the University of Southern California. He is now with NTT Transmission Systems Laboratories, Kanagawa, Japan.

Mr. Ito is a member of the Institute of Electronics, Information, and Communication Engineers of Japan and the Japan Society of Applied Physics.

Rafael Navarro is currently a Scientific Collaborator at the Institute of Optics, CSIC (National Council for Scientific Research), Madrid, Spain, since 1987. He received the M.Sc. and Ph.D. degrees in Physics from the University of Zaragoza, Spain in 1980 and 1984 respectively. His research interests are in the areas of physics of images and vision: firstly, image processing including restoration from blur, image reconstruction (phase retrieval) and applying models of human vision to image coding and analysis; and secondly, developing new methods to assess the image quality in physiological optics based on coherent optics and image processing techniques. Rafael Navarro is a member of the Optical Society of America.

Antonio Taberneo Galán received his M.S. degree in physics from the University Complutense of Madrid, Spain, in 1988. Since 1989 he has been working at the Instituto de Optica (Madrid), pursuing his Ph.D. degree. His current interests concern the understanding of the visual system, and its applications to image analysis, coding, and processing.

List of Reviewers

It is our great pleasure to list below the names of reviewers who have been responsible for maintaining the quality of this journal with their prompt, painstaking, and thoughtful comments. This list covers those who sent their reviews before August 31, 1991.

Aboulnasr, T.
Aburdene, M.F.
Ackermann, J.E.
Agathoklis, P.
Ansari, R.
Antoniou, A.
Aravena, J.L.
Barbour, D.K.
Basu, S.
Bates, R.H.T.
Bauer, P.
Beex, A.A.
Bello, M.
Berenstein, C.A.
Bhattacharyya, S.P.
Biemond, J.
Bisiacco, M.
Bose, N.K.
Boyd, S.P.
Brofferio, S.
Bruton, L.
Campbell, S.L.
Caratto, S.
Chaparro, L.F.
Chiasson, J.
Coleman, J.
Cooley, J.W.
Cybenko, G.
Dainty, J.C.
DeFacio, B.
Deprettre, E.F.A.
Dierieck, C.
Domash, L.
El-Jaroudi, A.
Emre, E.
Fahmy, M.M.
Fettweis, A.
Fiddy, M.A.
Fielding, K.
Fornasini, E.
Gallagher, N.C.
Garloff, J.
Gelfand, S.
Genin, Y.
Gerbrands, J.J.

Gharavi, H.
Gilge, M.
Giannakis, G.B.
Goodman, D.
Hall, E.
Henrichsen, D.
Hinamoto, T.
Hollot, C.V.
Horner, J.
Hurt, A.E.
Ingle, V.K.
Jagadish, H.V.
Jury, E.I.
Kaczorek, T.
Katbab, A.
Katsaggelos, A.K.
Kaufman, H.
Kaveh, M.
Kim, S.P.
Kim, K.D.
Kleihorst, R.P.
Kogan, J.
Krener, A.J.
Krishna, H.
Kummert, A.
Lagendijk, R.L.
Leonardi, R.
Levy, B.C.
Lewis, F.L.
Lin, Z.
Lin, F.C.
Longo, G.
Maitre, H.
Mangoubi, R.
Mansour, M.
Marchesini, G.
Mersereau, R.M.
Michel, A.N.
Misra, P.
Munson, D.C.
Musmann, H.G.
Nie, X.
Nieto-Vesperinas, M.
Paraskevopoulos, P.N.
Pitas, I.

Premaratne, K.
Prince, J.
Raghuramireddy, D.
Raghuveer, M.R.
Rajala, S.
Rajan, P.K.
Ramachandran, V.
Ramponi, G.
Rank, K.
Rao, K.R.
Rinaldi, S.
Rocha, P.
Roy, R.H.
Roytman, L.
Sarkar, T.
Schamel, G.
Sebek, M.
Shankar, S.
Shi, Y.Q.
Sicuranza, G.L.
Siljak, D.D.
Soh, C.B.
Sontag, E.
Stark, H.
Tayebati, P.
Tekalp, A.M.
Tempo, R.
Tits, A.L.
Tonge, G.J.
Trussell, J.
Tsypkin, Y.Z.
Unbehauen, R.
Valenzuela, H.M.
Wang, P.S.
Wang, P.W.
Warner, C.R.
Wendland, B.
Willems, J.C.
Willsky, A.S.
Woods, J.W.
Yagle, A.E.
Zakhor, A.
Zampieri, S.
Zeheb, E.

Table of Contents: Volume 2, 1991

Number 4

Multidimensional Systems and Signal Processing

Information for Authors

Authors are encouraged to submit high quality, original works which have not appeared, nor are under consideration, in other journals. Papers which have previously appeared in conference proceedings will also be considered, and this should be so indicated at the time of submission.

PROCESS FOR SUBMISSION

1. Authors should submit four hard copies of their final manuscript to the Editor for the author's region, at the following addresses:

North American Region
MARWAN SIMAAN
Dept. of Electrical Engineering
348 Benedum Hall
University of Pittsburgh
Pittsburgh, PA 15261
Phone: (0) (412) 624-9683
FAX No.: (0)(412)624-1108
E Mail: Simaan@ee.pitt.edu

European Region
JAN BIEMOND
Faculty of Electrical Engineering
Delft University of Technology
Mekelweg 4
P.O. Box 5031
2600 GA Delft
THE NETHERLANDS
Phone: (0) +31 15784695
FAX No.: +31 15783622

Regions Outside North America and Europe
NIRMAL K. BOSE
Dept. of Electrical Engineering
The Pennsylvania State University
121 Electrical Engineering East
University Park, PA 16802
Phone: (0) (814) 865-3912
FAX No.: (0)(814)865-7065
E Mail: nkb@psuecl.bitnet

2. Enclose with each manuscript, on a separate page, from five to ten index terms (key phrases).
3. Enclose originals for the illustrations in the style described below. Alternatively, good quality copies may be sent initially, with the originals ready to be sent immediately upon acceptance of paper. Also upon acceptance of the paper, authors must supply a photograph and brief biographical sketch.
4. Enclose a separate page giving your preferred address for correspondence and return of proofs. Please be sure to include your telephone number.
5. The refereeing is done by anonymous reviewers.
6. Papers exceeding 40 pages in length (including figures) may be returned to the author for shortening.
7. All papers should be written in English.

STYLE FOR MANUSCRIPT

1. Typeset, double or 1½ space; use one side of sheet only (laser printed, typewritten and good quality duplication acceptable).
2. Provide an informative 100- to 250-word abstract at the head of the manuscript. The abstracts are printed with the articles.
3. Provide a separate double-spaced sheet listing all footnotes, beginning with "Affiliation of author" and continuing with numbered references. Acknowledgement of financial support may be given if appropriate.
4. References should be numbered and listed in the order of citation in a separate section at the end of the paper, with citations by numbers in square brackets (e.g., [1], [2], etc.).
   - Style for papers: Author (initials and surname), "Paper Title," *Journal Title*, volume number, date, page numbers. For example: J. Fox, G. Surace, and P.A. Thomas, "A Self-Testing 2-μm CMOS Chip Set for FFT Applications," *IEEE Journal of Solid-State Circuits*, vol. SC-22, 1987, pp. 15–19.
   - Style for books: Author, *Book Title*, location, publisher, date, chapter or page numbers (if desired). For example: P.M. Kogge, *The Architecture of Pipelined Computers*, New York: McGraw-Hill Book Co., 1981.
5. Provide a separate sheet listing all figure captions, in proper style for the typesetter, e.g., "Fig. 3. Examples of the zero crossings of the second derivative of the (a) Gaussian and (b) sine filter for the same input function."
6. Type or mark mathematical copy exactly as it should appear in print. Journal style for letter symbols is as follows: variables, italic type (indicated by underline); constants, roman text type; matrices and vectors, boldface type (indicated by wavy underline). In word-processor manuscripts, use appropriate typeface. It will be assumed that letters in displayed equations are to be set in italic type unless you mark them otherwise. All letter symbols in text discussion must be marked if they should be italic or boldface. Indicate best breaks for equations in case they will not fit on one line.

STYLE FOR ILLUSTRATIONS

1. Originals for illustrations should be sharp, noise-free, and of good contrast. We regret that we cannot provide drafting or art service.
2. Line drawings should be in laser printer output or in India ink on paper, or board. Use 8½ by 11-inch size sheets if possible, to simplify handling of the manuscript.
3. Each figure should be mentioned in the text and numbered consecutively using Arabic numerals. Specify the desired location of each figure in the text, but place the figure itself on a separate page following the text.
4. Number each table consecutively using Arabic numerals. Please label any material that can be typeset as a table, reserving the term "figure" for material that has been drawn. Specify the desired location of each table in the text, but place the table itself on a separate page following the text. Type a brief title above each table.
5. All lettering should be large enough to permit legible reduction.
6. Photographs should be glossy prints, of good contrast and gradation, and any reasonable size.
7. Number each original on the back, or at the bottom of the front.
8. Include a separate page listing the captions for all figures.

PROOFING

Page proofs for articles to be included in a journal issue will be sent to the first named author for proofing. The proofread copy should be received back by the Publisher within 72 hours.

COPYRIGHT

Upon acceptance of an article, authors will be required to sign a copyright form transferring the copyright from the authors or their employers to the publisher.

REPRINTS

First-named authors will be entitled to 50 free reprints of their paper.